T0230444

# Lecture Notes in Computational Vision and Biomechanics

## Volume 24

**Series editors**

João Manuel R.S. Tavares, Porto, Portugal
Renato Natal Jorge, Porto, Portugal

**Editorial Advisory Board**

Alejandro Frangi, Sheffield, UK
Chandrajit Bajaj, Austin, USA
Eugenio Oñate, Barcelona, Spain
Francisco Perales, Palma de Mallorca, Spain
Gerhard A. Holzapfel, Stockholm, Sweden
J. Paulo Vilas-Boas, Porto, Portugal
Jeffrey A. Weiss, Salt Lake City, USA
John Middleton, Cardiff, UK
Jose M. García Aznar, Zaragoza, Spain
Perumal Nithiarasu, Swansea, UK
Kumar K. Tamma, Minneapolis, USA
Laurent Cohen, Paris, France
Manuel Doblaré, Zaragoza, Spain
Patrick J. Prendergast, Dublin, Ireland
Rainald Löhner, Fairfax, USA
Roger Kamm, Cambridge, USA
Shuo Li, London, Canada
Thomas J.R. Hughes, Austin, USA
Yongjie Zhang, Pittsburgh, USA

The research related to the analysis of living structures (Biomechanics) has been a source of recent research in several distinct areas of science, for example, Mathematics, Mechanical Engineering, Physics, Informatics, Medicine and Sport. However, for its successful achievement, numerous research topics should be considered, such as image processing and analysis, geometric and numerical modelling, biomechanics, experimental analysis, mechanobiology and enhanced visualization, and their application to real cases must be developed and more investigation is needed. Additionally, enhanced hardware solutions and less invasive devices are demanded.

On the other hand, Image Analysis (Computational Vision) is used for the extraction of high level information from static images or dynamic image sequences. Examples of applications involving image analysis can be the study of motion of structures from image sequences, shape reconstruction from images, and medical diagnosis. As a multidisciplinary area, Computational Vision considers techniques and methods from other disciplines, such as Artificial Intelligence, Signal Processing, Mathematics, Physics and Informatics. Despite the many research projects in this area, more robust and efficient methods of Computational Imaging are still demanded in many application domains in Medicine, and their validation in real scenarios is matter of urgency.

These two important and predominant branches of Science are increasingly considered to be strongly connected and related. Hence, the main goal of the LNCV&B book series consists of the provision of a comprehensive forum for discussion on the current state-of-the-art in these fields by emphasizing their connection. The book series covers (but is not limited to):

- Applications of Computational Vision and Biomechanics
- Biometrics and Biomedical Pattern Analysis
- Cellular Imaging and Cellular Mechanics
- Clinical Biomechanics
- Computational Bioimaging and Visualization
- Computational Biology in Biomedical Imaging
- Development of Biomechanical Devices
- Device and Technique Development for Biomedical Imaging
- Digital Geometry Algorithms for Computational Vision and Visualization
- Experimental Biomechanics
- Gait & Posture Mechanics
- Multiscale Analysis in Biomechanics
- Neuromuscular Biomechanics
- Numerical Methods for Living Tissues
- Numerical Simulation
- Software Development on Computational Vision and Biomechanics
- Grid and High Performance Computing for Computational Vision and Biomechanics
- Image-based Geometric Modeling and Mesh Generation
- Image Processing and Analysis
- Image Processing and Visualization in Biofluids
- Image Understanding
- Material Models
- Mechanobiology
- Medical Image Analysis
- Molecular Mechanics
- Multi-Modal Image Systems
- Multiscale Biosensors in Biomedical Imaging
- Multiscale Devices and Biomems for Biomedical Imaging
- Musculoskeletal Biomechanics
- Sport Biomechanics
- Virtual Reality in Biomechanics
- Vision Systems

More information about this series at http://www.springer.com/series/8910

Vytautas Ostasevicius · Giedrius Janusas
Arvydas Palevicius · Rimvydas Gaidys
Vytautas Jurenas

# Biomechanical Microsystems

## Design, Processing and Applications

 Springer

Vytautas Ostasevicius
Institute of Mechatronics
Kaunas University of Technology
Kaunas
Lithuania

Giedrius Janusas
Faculty of Mechanical Engineering
 and Design
Kaunas University of Technology
Kaunas
Lithuania

Arvydas Palevicius
Faculty of Mechanical Engineering
 and Design
Kaunas University of Technology
Kaunas
Lithuania

Rimvydas Gaidys
Faculty of Mechanical Engineering
 and Design
Kaunas University of Technology
Kaunas
Lithuania

Vytautas Jurenas
Institute of Mechatronics
Kaunas University of Technology
Kaunas
Lithuania

ISSN 2212-9391          ISSN 2212-9413   (electronic)
Lecture Notes in Computational Vision and Biomechanics
ISBN 978-3-319-85500-4          ISBN 978-3-319-54849-4   (eBook)
DOI 10.1007/978-3-319-54849-4

© Springer International Publishing AG 2017
Softcover reprint of the hardcover 1st edition 2017
This work is subject to copyright. All rights are reserved by the Publisher, whether the whole or part
of the material is concerned, specifically the rights of translation, reprinting, reuse of illustrations,
recitation, broadcasting, reproduction on microfilms or in any other physical way, and transmission
or information storage and retrieval, electronic adaptation, computer software, or by similar or dissimilar
methodology now known or hereafter developed.
The use of general descriptive names, registered names, trademarks, service marks, etc. in this
publication does not imply, even in the absence of a specific statement, that such names are exempt from
the relevant protective laws and regulations and therefore free for general use.
The publisher, the authors and the editors are safe to assume that the advice and information in this
book are believed to be true and accurate at the date of publication. Neither the publisher nor the
authors or the editors give a warranty, express or implied, with respect to the material contained herein or
for any errors or omissions that may have been made. The publisher remains neutral with regard to
jurisdictional claims in published maps and institutional affiliations.

Printed on acid-free paper

This Springer imprint is published by Springer Nature
The registered company is Springer International Publishing AG
The registered company address is: Gewerbestrasse 11, 6330 Cham, Switzerland

# Preface

Over the past few decades, raising public awareness of the health, physical activity sensing creates new demands for smart sensor technology and monitoring devices capable of feeling, to classify and to provide feedback to users of health and physical activity in common, accurate and reliable fashion. Monitoring and accurately quantify the physical activity of users with devices inertial unit on the basis of measurements, for example, also proved an important role in health management of patients with chronic diseases. The purpose of this book will focus on MEMS-MOEMS sensor technology developed in the past few years, describing the scientific achievements on health and physical activity in addition to the smart systems manufacturing and integration.

Research monograph focuses on the dynamic aspects of microsystems, presenting a detailed numerical analysis of the different types of microsystems, which are studied from a mechanical point of view, thus focusing on the complex process and the internal dynamics of elastic structures such as natural vibration modes and their beneficial use. Computational models proposed to take into account the interaction between micro devices and parts of the human body. The adequacy of these models confirmed using experimental precision measurement methods. Some important issues such as the MEMS assisted for human obesity prevention or MOEMS based radial pulse measurements are presented in this book.

This monograph is of use to researchers, practitioners and manufacturers in the field of biomechanical microsystems engineering and may be used by Ph.D. students for advanced courses as additional material.

This research was funded by grants (No. SEN-10/15, No. MIP-026/2014 and No. MIP-081/2015) from the Research Council of Lithuania.

Kaunas, Lithuania
December 2016

Vytautas Ostasevicius
Giedrius Janusas
Arvydas Palevicius
Rimvydas Gaidys
Vytautas Jurenas

# Contents

# Chapter 1
# Introduction

**Abstract** This chapter introduces to the field of microelectromechanical systems for biomechanical applications. The areas of application of implatable, non-invasive biomedical sensors as well as the importance of blood pressure and radial pulse diagnosis using MEMS are discussed.

## 1.1 Implantable Biomedical Sensors

Strategy for miniaturization and integration gains outstanding value for the health sector for economic reasons, as well as on the fundamental performance improvements. Micro devices meanwhile are applied to medical diagnosis and therapy, particularly for the development of biochemical drugs. Pacemakers, vascular stents, hearing aids, cochlea implants or microelectrodes to stimulate nerves should be mentioned, which is much more efficient to maintain the function of human organs, than the former devices manufactured by standard macroscopic manufacturing methods. Instruments for minimally invasive diagnostics and therapy, as micro-endoscopes and micro-catheter systems are small, but powerful tools to surgeons and help reduce the time and pain of operations.

Micro acceleration sensors are integrated in the artificial knee joint with a sharp improvement for patient mobility. Currently, the main developments in the more advanced micro-technology health care products are aimed at the replacement of complex organic functions or even complete organs. Extensive progress has been made in drug delivery systems based on a micro-technology. A very promising device with a high market potential uses micro nozzles for the generation of inhalable drugs aerosol with a defined droplet size that is optimal for the absorption by lung. Commercially available as implantable devices with a micro-capillary loops for controlled release of analgesics and drugs for cancer. Other drug delivery systems will be equipped with a programmable micro pumps for precise control of the release. There is no doubt that the desire for health provides an interesting basis for progress and gains in medical technology companies, which are able to make use of the inherent advantages of micro-technology. There are attractive

© Springer International Publishing AG 2017
V. Ostasevicius et al., *Biomechanical Microsystems*, Lecture Notes
in Computational Vision and Biomechanics 24,
DOI 10.1007/978-3-319-54849-4_1

opportunities for business, particularly for small and medium-sized enterprises, while collaborating with physicians and specialists in micro-fabrication.

## 1.2  Non-invasive Sensors and Application Areas

Nowadays non-invasive methods used for human blood system investigations mainly are based on applications of ultrasound, laser or infrared light.

Micro spectrometers are used for color measurement as a diagnostic tool for non-invasive measurement of bilirubin. Dentists use a micro-spectrometers, to accurately determine the color of teeth for dentures. Well known world company Futrex is the leader in the devices for the investigation of fat of human body, once again with non invasive methods. The applications of such devices are very wide, namely from university students to military soldiers and patients of rehabilitation hospitals.

The working principle is that during operation Futrex sends safe, a near infrared light beam to the triceps at certain wavelengths that will be absorbed by fat and reflected by muscle. Absorption of light is measured to determine the body fat. The test results are traceable to underwater weighing are displayed instantly for your customers to see. Using the same background, namely reflection of near infrared light beam one can easily obtain the concentration of some chemical element of the blood. Also glucose and cholesterol concentration measurement method, and non-invasive "in vitro" using nuclear magnetic resonance spectroscopy is widely used. Measurement comprises a ratio formed by dividing the area of resonance of the desired analyte, e.g., glucose or cholesterol, in aqueous resonance spectrum of blood or tissue. In an environment in vivo, the spectrum is obtained by or in connection with the blood pulsation or by selecting the slice gradient, for example, used in magnetic resonance imaging. This measurement is then correlated with conventional serum concentration of the analyte.

Skin test for cholesterol detects early signs of heart disease. The basic advantage of such method is that the concentration of cholesterol in blood can be measured in less than 5 min, meanwhile traditional invasive cholesterol measurements usually takes even up to 24 h. As this method is non-invasive, it is obvious that no needles are used for this type of analysis that is why it is widely used for the infants and small kids—without giving them stress.

Nowadays it is possible to get only tail cuff devices for blood pressure and pulse analysis. The tail cuff can receive only the systolic and diastolic blood pressure levels, instead of the continuous blood pressure and pulse waveforms detailed signatures that are desirable for advanced biomedical research. So far commercial technologies are inadequate for long-term real-time monitoring of blood pressure and heart rate analysis.

## 1.3   Importance of Blood Pressure and Radial Pulse Diagnosis

Blood pressure measurement is one of the most important medical measurements. Blood pressure is the first assessment of blood flow and still is the easiest parameter to measure. Thus, it is a convenient measure of the patient's health, as well as one of the most important vital functions. The importance of measuring blood pressure is that it is firmly connected, via the impedance of human organs, with the physiology of the human body, and nearly all physiological processes that are carried in blood pressure signals, either arterial or venous. Additionally, some features of arterial blood pressure, such as mean arterial pressure, systolic and diastolic pressure are epidemiologically associated with numerous circulatory system diseases such as hypertension, obesity, epilepsy and cancer, myocardial infarction, stroke, congestive heart failure valve, atherosclerosis, and as it regards the problems associated with diabetes and renal diseases.

The biggest advantage in the field of MOEMS-MEMS-devices is that they open up entirely new possibilities for more accurate, permanent blood pressure in real-time measurements. In the 21st century, numerous technologies were developed and introduced to the market, such as implantable pacemakers, defibrillators, neural-muscular stimulators, and new classes of biomechanical prostheses (e.g., skeletal prosthesis), vascular grafts and heart assist devices. These developments in the field of biocompatibility and implant surgery have put new tools in the hands of biomedical engineers working on new arterial blood pressure measurement methods. Growing confidence in implantable devices can shift the point of gravity of the non-invasive, less accurate diagnostic tools towards miniature sensors implanted with high accuracy.

In addition, the newest trends in medicine tend to employ close loop feedback systems (e.g. in drug delivery devices, pacemakers, implantable defibrillators, etc.) that require real-time reading of the control signal. Therefore, future anti-hypertension drug delivery systems most likely will be based on real-time arterial blood pressure measurements. The same is true for heart assist devices.

All those facts call for long-term, real-time "waveform-capable" methods. Arguably the best solution to all those problems will be the miniature implantable blood pressure sensor capable of constant measuring of the real-time blood pressure waveform. Thus, the future of blood pressure measurements lies in the merger of arterial tonometry and an implantable device that is an implantable arterial tonometer.

Arterial tonometry preserves the arterial wall, and if it is applied directly to the artery, it exhibits superb accuracy. This kind of device has the potential to provide high accuracy and real-time measurements with low biocompatibility risks. The risks from insufficient diagnostics and early prophylactics in the field of hypertension could be significantly higher than potential risks of implantation of the miniature arterial tonometer.

## 1.4  Obesity as the 21st Century Plague

According to the World Health Organization Regional Office for Europe, obesity is one of the most important public health challenges of the 21st century. It tripled from the 1980s in the European region and the number of victims continues to grow at an alarming rate, especially among children, who are our future. Obesity causes various physical disabilities and physiological problems and excess weight dramatically increases a person's risk of developing a number of non-communicable diseases like cardiovascular disease, cancer and diabetes. The risk of having multiple such diseases (co-morbidity) also increases with body weight. Obesity is already responsible for 2–8% of health costs and they can be as high as 10.4 billion euros—and 10–13% of deaths in different parts of European Region.

Relationships between physical activity and mortality, cardiovascular diseases and 2nd type diabetes spread are also well established. Scientific research results provoked a new term—sitting lifestyle death syndrome—whose indications are lower bone density, higher sugar levels in blood and urine, obesity, bad aerobic stamina, tachycardia in calm state, all of the mentioned being the source of disturbance for body organs and systems. Unfortunately, as a result, today's generation of adolescents who are less than 25 years old, and make nearly half of the world's population, face far more complex challenges to their health and development than their parents did. It is thus important for people to become motivated regarding the importance of physical activity and nutrition for their health and wellbeing and be encouraged to practice a healthy lifestyle. Perception of healthcare should be based upon the fact that health starts with prevention, by carefully considering nutrition patterns, activity schedules and predisposition to several diseases, thus about our complete lifestyle.

There are number of tools that act as a prevention measure to boost person's motivation for physical activity and its levels. These tools are produced by well-known names as Polar, Suunto, Zephyr and others, and are available to buy in different consumer markets. However they are more oriented to monitor active physical activities like running, cycling, etc. but not every day physical activities people do on a regular basis with only few exceptions.

Devices that are oriented for sports tracking usually have a lot of features that are beneficial to have in sports but are not mandatory for daily monitoring. Such are heart rate thresholds, GPS, barometer, integrated fitness tests, etc. These sports oriented devices are usually highly profiled to fit the needs of different sports. The most popular sports are running, cycling, diving so profiled devices have specific features that are useful in that particular sport.

Most of these physical activity measurement and evaluation tools use acceleration measurements as their input data. Some of the devices take advantage of other input dimensions like heart rate data for better accuracy of evaluation and additional benefits but that also require additional hardware which is inconvenient to wear on a daily basis.

There is a class of devices that use accelerometer measurements as a primary data source and are daily physical monitoring oriented. The placement of these devices is not restricted (meaning there is some freedom the user can have while wearing the device) thus, measuring errors that occur due to the body rheology, clothing, etc. cannot be eliminated. These errors are bypassed using statistical indirect methods to evaluate physical activity levels and minimize errors impact toward the result. For example: Actismile is worn on a strap around the neck (which allows unrestrictive movements during more intensive physical activities) and uses only vertical acceleration data to evaluate physical activity levels by classifying human activity into 4 different classes. This means that the device can provide only rough estimates of physical activity levels.

Actigraphy is worn on the waist or arm, and uses a 3-axis MEMS accelerometer for accelerations recording time changes in magnitude in a range from $\pm 6$ g's. The output of the accelerometer is digitized into twelve bits (12) analog-to-digital converter in the range from 30 to 100 Hz. Using software Actilife, the raw data collected by the device can be viewed directly or further processed. Each sample was then added by a user-defined time interval is called the "epoch". This again provides integrated results with initial acceleration data put through digital filter that is band-limited from 0.25 to 2.5 Hz. No input corrections are present whatsoever and with specified digital signal filters applied only the highly filtered accelerations are used in further calculations.

If the placement of mentioned and similar devices would be more restricted allowing the implementation of error reduction algorithms, the processing of the data base that is the input for further analysis and evaluation would likely yield better (more accurate) evaluation/analysis results.

# Chapter 2
# Development of Microsystems Multi Physics Investigation Methods

**Abstract** The theoretical and experimental methods for the investigation of microsystems multi physic processes are presented. The FEM method for the analysis of MEMS in digital environment in combination with experimental data from holographic interferometry is developed. Numerical–experimental method for evaluation of geometrical parameters and their usage for characterization of microstructures is presented. In order to optimise hot imprint method in polycarbonate, an elasto-plastic material model for simulation of microstructures hot imprint method is developed.

## 2.1 Application of Time Averaged Holography for Micro-Electro-Mechanical System Performing Non-linear Oscillations

To perform the analysis of the links of MEMS systems time average laser holography [1] may be applied. Time average laser holography is a non-destructive full field technique, which may be used of investigation and analysis of dynamics of vibrating amplitudes and static displacements of deformable surface of MEMS components [2]. In combination with optical and digital holography numerous numerical methods [3] used for interpretation of patterns of fringes of holographic interferograms of analysed MEMS. Sometimes because of nonlinearities of MEMS links interpretation and analysis of holographic interferogram needs additional numerical investigations. An example of synergy of this methodology is presented in this chapter. The sequency of manufacturing technology of a micro-electromechanical switch used for the analysis follows.

The manufacturing of the MEM switch begins with the patterning and reactive ion etching of silicon using $SF_6/N_2$ gas chemistry in the cantilever support area fabricating microstructures to increase the cantilever bond strength either durability of the device. After treatment of the substrate in the $O_2/N_2$ gases mixture plasma chrome layer of about 30 nm thickness and gold layer of about 200 nm thickness were deposited. Patterning of the source, gate and drain electrodes were performed

© Springer International Publishing AG 2017
V. Ostasevicius et al., *Biomechanical Microsystems*, Lecture Notes
in Computational Vision and Biomechanics 24,
DOI 10.1007/978-3-319-54849-4_2

using lift-off lithography. Electron beam evaporation was performed to deposit a sacrificial copper layer with thickness of about 3000 nm. Copper layer covered the whole area of the substrate. Patterning of the copper layer was performed in two steps. First of all, the copper layer was partially etched (etchant: $H_2SO_4:CrO_3:H_2O$) to define the contact tips for the cantilever and etching duration directly determined the spacing between tip's top and drain electrode. Next, the copper layer was etched away to uncover the source cantilever support area. The next step was photo resist patterning on the top of the sacrificial layer to define the mask for the cantilever sector and lift-off lithography of the evaporated gold layer with thickness of about 200 nm was performed. Afterwards, the photoresist was spun and patterned once again in the same sector and thick nickel layer was electroplated (sulfamate electrolyte: $Ni(NH_2SO_3)_2:4H_2O$) fabricating cantilever structure. Finally, the sacrificial layer was removed away using the same wet copper etchant to release the free-standing cantilever. The general view of MEMS cantilever is presented in Fig. 2.1.

The holographic interferogram of MEMS cantilever are presented in Fig. 2.2. The methodology of recording holographic interferogram is described in [4]. During registration holographic interferogram at first MEMS cantilever was excited acoustically (Fig. 2.2a). In Fig. 2.2b holographic interferogram is registrated of cantilever excited by oscillating charge of the drain electrode.

The ordinary fringe counting techniques are applied for the reconstruction of the field of vibration amplitudes in case of acoustically excited cantilever because the sinusoidal periodical excitation was used and the methods for interpretation of time-average holographic interferograms are discussed well.

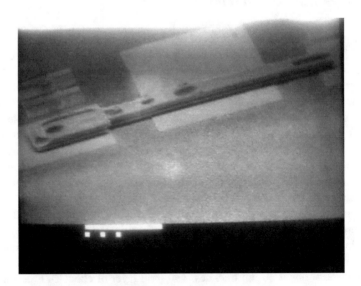

**Fig. 2.1** Microscopic photo of MEMS cantilever

(a)

(b)

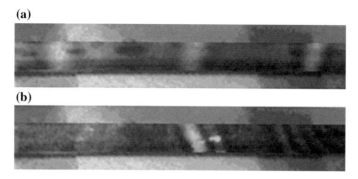

**Fig. 2.2** Holographic interferograms of cantilever: **a** holographic interferogram of cantilever acoustically excited; **b** holographic interferogram of cantilever excited by oscillating charge

The interpretation of holographic interferogram presented in Fig. 2.2b is much more complicated. The knowing the fact that cantilever excited by oscillating charge assist as to evaluate holographic interferogram taking into account the nonlinearities of vibrating surface of cantilever. Because of chaotic processes of vibrating cantilever the time exposure of recording holographic interferogram should be managed because longer exposure times produce dark images when the cantilever is excited by oscillating charge and the pattern of fringes is sensitive to exposure time and the quality of the holographic images are pure.

The interpretation of pattern of fringes in Fig. 2.2b is much more complicated—one white and several dark fringes are distributed on the surface of cantilever and it is quite difficult to understand the dynamics of the cantilever. Moreover, longer exposure times produce dark images when the cantilever is excited by oscillating charge and the pattern of fringes is sensitive to exposure time. Developing hybrid numerical–experimental models of analysed dynamical systems in this case let as to analyse generated patterns of fringes in holograms of MEMS cantilevers excited by oscillating charge originated in order to get much more characteristic about the motion of the objects.

This approach when simulation of the dynamic is used as well as optical processes taking place in the analysed systems could help understanding experimental results.

## 2.1.1  Phenomenological Model of MEMS Cantilever

To achieve this goal in analysing complex MEMS cantilever motion the development of simple one-dimensional phenomenological model is presented in Fig. 2.3.

Governing equation of motion of the system presented in Fig. 2.3 takes the following form:

**Fig. 2.3** One degree of
freedom phenomenological
model of MEMS cantilever

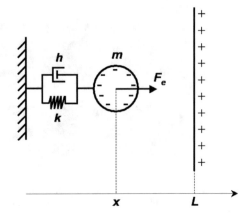

$$m\ddot{x} + h\dot{x} + kx = F_e(x) \tag{2.1}$$

where $m$, $h$, $k$—mass, viscous damping and stiffness coefficients; $x$—co-ordinate; $F_e$—electrostatic force; top dots denote full derivatives by time $t$. It is assumed that the charged contact plane is motionlessly fixed at co-ordinate $L$ (Fig. 2.3). Mass $m$ is negatively charged, while the charge $Q$ of the contact plane varies harmonically in time:

$$Q = q\sin(\omega t) \tag{2.2}$$

where $q$—maximum charge of the contact plane; $\omega$—frequency of charge oscillation. Then the electrostatic force $F_e$ acting to mass $m$ is

$$F_e = C\frac{Q}{L - x} \tag{2.3}$$

where constant $C$ depends from the charge of mass $m$, density of air, etc.

The charge of the contact plane because of harmonic oscillation provides a strongly nonlinear response. That is a result because governing equation of motion is non-linear. The non-linearity of the vibration of the mass $m$ occur when the frequency of charge oscillation is around the natural frequency of the cantilever (Fig. 2.4). Usually the frequency of charge oscillation very rarely reaches the fundamental frequency of the MEMS cantilever only due to the fact that it is very high and the excitation frequencies are of magnitude lower than fundamental frequencies.

The developed phenomenological cantilever model is analysed when the excitation frequency is much lower that the fundamental frequency and the differential equation turns to be stiff and special care is required applying direct time marching integration techniques. First the system is integrated until the transient processes cease down. Then the attractor in phase plane $x - \dot{x}$ is drawn. Array of attractors as shown in Fig. 2.5 is built of repetition of such procedure at different values of

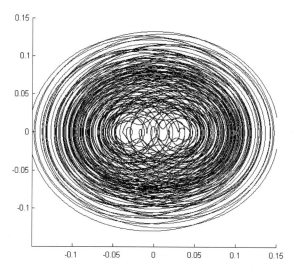

**Fig. 2.4** Chaotic motion of mass m in phase plane $x - \dot{x}$ at $m = 1$; $k = 1$; $h = 0.003$; $\omega = 0.73$; $q = 0.1$; $L = 2$

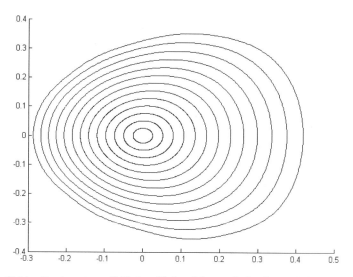

**Fig. 2.5** Stable attractors at $m = 0.25$; $k = 20$; $h = 0.1$; $\omega = 1$; $L = 2$; $q = 1, 2, …, 13$

maximum charge $q$. The motion of the mass $m$ is almost linear at small excitation. The form of the stable attractor gradually deforms at increasing excitation what is a natural result for a non-linear system.

When we use holographic interferometry method how the experimental results could be interpreted if the analysed object would oscillate not harmonically, but as shown in Fig. 2.5.

For simplicity we deal with one-dimensional system. Then the intensity of illumination $I$ in the hologram plane will be:

$$I = \lim_{T \to \infty} \frac{1}{T^2} \left| \int_0^T \exp\left(j\frac{2\pi}{\lambda}\zeta(t)\right) dt \right|^2 \tag{2.4}$$

where $T$—exposure time; $\lambda$—laser wavelength; $\zeta$—scalar time process; $j$—imaginary unit. When $\zeta(t) = a\sin(\omega t + \varphi)$ where $a$, $\omega$, $\varphi$—amplitude, angular frequency and phase of oscillations, the intensity of illumination takes the form:

$$I = \left| J_0\left(\frac{2\pi}{\lambda}a\right) \right|^2 = \lim_{T \to \infty} \frac{1}{T^2} \left( \int_0^T \cos\left(\frac{2\pi}{\lambda}a\sin(\omega t + \varphi)\right) dt \right)^2$$

$$\approx \left( \frac{1}{m}\sum_{i=1}^m \cos\left(\frac{2\pi}{\lambda}a\sin\left(\frac{2\pi}{m}(i-1)\right)\right) \right)^2 \tag{2.5}$$

where $J_0$—zero order Bessel function of the first kind. It can be noted that $\lim_{T \to \infty} \int_0^T \sin\left(\frac{2\pi}{\lambda}a\sin(\omega t + \varphi)\right) dt = 0$ due to evenness of the sine function, and that the angular frequency and phase have no effect to the intensity of illumination. The second approximate equality builds the ground for numerical modelling of the relationships governing the formation of interference fringes.

If $\zeta(t)$ is not a harmonic process the intensity of illumination can be numerically reconstructed from Eq. (2.4), but the calculation is more complex than in Eq. (2.5) due to the fact the integral $\lim_{T \to \infty} \int_0^T \sin\left(\frac{2\pi}{\lambda}\zeta(t)\right) dt$ does not converge to zero. If $\zeta(t)$ is a periodic process and $T_p$ is the time length if the period, the approximate numerical calculation scheme takes the following form:

$$I \approx \left( \frac{1}{m}\sum_{i=1}^m \cos\left(\frac{2\pi}{\lambda}\zeta\left(t_0 + \frac{T_p}{m}(i-1)\right)\right) \right)^2$$

$$+ \left( \frac{1}{m}\sum_{i=1}^m \sin\left(\frac{2\pi}{\lambda}\zeta\left(t_0 + \frac{T_p}{m}(i-1)\right)\right) \right)^2 \tag{2.6}$$

where $t_0$—arbitrary selected time moment. If $\zeta(t)$ is a process characterising time history of a dynamical system setting to a stable limit cycle type attractor, time moment $t_0$ must be selected large enough so that the transient processed are ceased. Such calculations are performed for an array of attractors shown in Fig. 2.6. One hundred separate solutions of Eq. (2.1) are analysed at intermittent values of $q$ in the range from 0 to 13. The produced intensities of illumination are presented in Fig. 2.6 ($x$ axis denotes 100 separate problems).

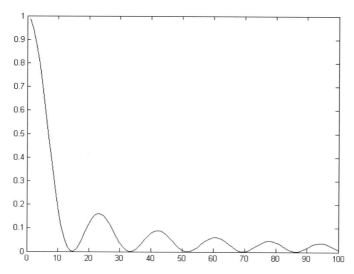

**Fig. 2.6** Relationship between intensities of illumination and parameter $q$

Remarkable is the fact that the relationship between the intensity of illumination and parameter $q$ holds the same character as the square of Bessel function in Eq. (2.5). In other words, the inverse problem of the reconstruction of dynamic displacements does not have a unique solution. Interference fringes can be formed when the analysed object performs harmonic vibrations. Very similar interference fringes will be formed when the object will perform non-linear periodic oscillations. Thus though time average laser holography is a very attractive technique for analysis of MEMS cantilever vibrations, the interpretation of experimentally produced interference fringes is rather complex procedure if one cannot be sure if the vibrations are harmonic. This effect is illustrated in Fig. 2.7. The presented non-linear periodic oscillation and harmonic vibration will both generate the same intensity of illumination corresponding to the centre of the sixth interference fringe. It can be noted that time average laser holography is insensitive to static shifts of harmonic oscillations. That follows from the property of Bessel function:

$$\lim_{T \to \infty} \frac{1}{T^2} \left| \int_0^T \exp\left( j\frac{2\pi}{\lambda}(a\sin(\omega t + \varphi) + C) \right) dt \right|^2$$

$$= \lim_{T \to \infty} \frac{1}{T^2} \left| \int_0^T \exp\left( j\frac{2\pi}{\lambda}(a\sin(\omega t + \varphi)) \right) dt \right|^2 = \left( J_0\left( \frac{2\pi}{\lambda}a \right) \right)$$

(2.7)

where $C$—constant. Results presented in Fig. 2.7 are remarkable not for the difference between the averages of non-linear and harmonic vibrations. Interesting is the fact that two different trajectories generate same intensity of illumination.

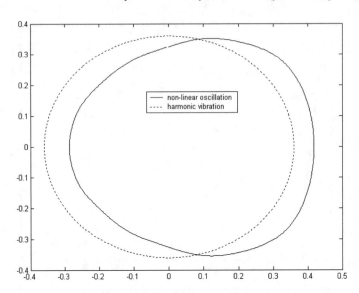

**Fig. 2.7** Two different trajectories generating same intensity of illumination

When we have chaotic oscillations the illumination of intensity can be calculated for certain stochastic time series approximating process $\zeta(t)$. If a time series $\zeta_i$ is normally distributed with variance $\sigma^2$ then, the decay of intensity of illumination can be calculated as follows:

$$I \approx \lim_{m\to\infty} \left( \left( \frac{1}{m} \sum_{i=1}^{m} \cos\left(\frac{2\pi}{\lambda}\zeta_i\right) \right)^2 + \left( \frac{1}{m} \sum_{i=1}^{m} \sin\left(\frac{2\pi}{\lambda}\zeta_i\right) \right)^2 \right)$$

$$= \lim_{m\to\infty} \left( \frac{1}{m} \sum_{i=1}^{m} \sum_{k=0}^{+\infty} \frac{(-1)^k \left(\frac{2\pi}{\lambda}\zeta_i\right)^{2k}}{(2k)!} \right)^2 = \left( \sum_{k=0}^{+\infty} \frac{(-1)^k \left(\frac{2\pi}{\lambda}\right)^{2k}}{(2k)!} \lim_{m\to\infty} \sum_{i=1}^{m} \frac{(\zeta_i)^{2k}}{m} \right)^2$$

$$= \left( \sum_{k=0}^{+\infty} \frac{(-1)^k \left(\frac{2\pi}{\lambda}\right)^{2k}}{(2k)!} \cdot (2k-1)!! \cdot \sigma^{2k} \right)^2 = \left( \sum_{k=0}^{+\infty} \frac{(-1)^k \left(\frac{2\pi}{\lambda}\sigma\right)^{2k}}{(2k)!!} \right)^2$$

$$= \left( \sum_{k=0}^{+\infty} \frac{(-1)^k \left(\frac{2\pi}{\lambda}\sigma\right)^{2k}}{2^k k!} \right)^2 = \left( \sum_{k=0}^{+\infty} \frac{(-1)^k}{k!} \left( \frac{1}{2} \left(\frac{2\pi}{\lambda}\sigma\right)^2 \right)^k \right)^2 = \exp^2\left( -\frac{1}{2}\left(\frac{2\pi}{\lambda}\sigma\right)^2 \right)$$

$$\tag{2.8}$$

The following identities are used in Eq. (2.8). If $\zeta \sim N(0, \sigma^2)$, then

$$E\zeta^{2k-1} \equiv 0, \quad k = 1, 2, 3, \ldots;$$
$$E\zeta^{2k} \equiv 1 \cdot 3 \cdot \cdots \cdot (2k-1)\sigma^{2k} = (2k-1)!!\sigma^{2k}, \quad k = 1, 2, 3, \ldots.$$

$$\tag{2.9}$$

It can be noted, that in this case no interference fringes will be formed at all—the intensity of illumination will gradually decrease at increasing variance $\sigma^2$.

This example shows an ill-posed inverse problem. The problem of interpretation of motion from the structure of the field of interference fringes has solution only if the vibration of the analysed system is harmonic. When the oscillations are non-linear (what is likely in MEMS cantilever dynamics) the interpretation of pattern of fringes is rather complicated.

## 2.1.2   FEM Analysis of MEMS Cantilever Performing Chaotic Oscillations

Virtual numerical environment is used for MEMS cantilever by FEM techniques with simulation of optical formation of optical holographic interferogram [5]. The first eight eigenmodes are presented in Fig. 2.8.

Complex dynamic response of the tip of the cantilever is got in case simulation of the dynamics of cantilever under oscillating charge excitation. The displacement of the tip of the cantilever is presented in Fig. 2.9. Various time exposures are used to generate holographic interferograms of the cantilever and the results are presented in Fig. 2.10.

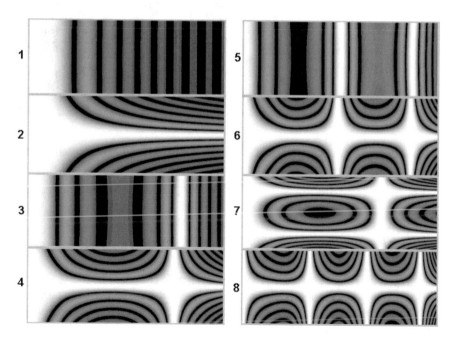

**Fig. 2.8** Time average holographic interferograms of the first eight eigenmodes of MEMS cantilever

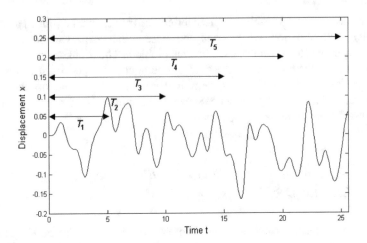

**Fig. 2.9** Chaotic dynamics of the tip of MEMS cantilever and times of exposure $T_1$–$T_5$

**Fig. 2.10** Time average
interferograms of the MEMS
cantilever at different times of
exposure

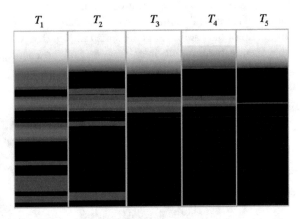

The complexity of the dynamical processes taking place in the analysed MEMS systems is illustrated by presented numerical results which validate the theoretical predictions and help to explain character of MEMS motion.

### 2.1.3 The Structure of Digital Data Processing

Holographic interferometry data processing system is presented in Fig. 2.11.

Box 1 represents the initial data of the analysis—that is, holographic interferogram holding information about the measured object and interference bands which denote the time variance of the surface of the analysed body [6]—Fig. 2.12.

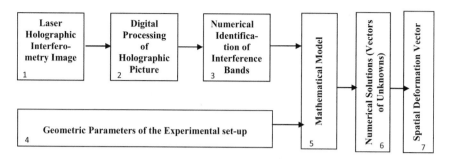

Fig. 2.11  Structure of data processing system

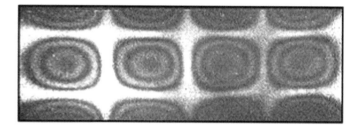

Fig. 2.12  Laser holographic interferometry image of a vibrating plate

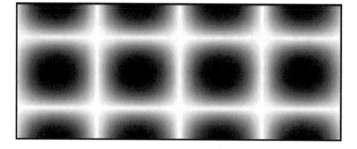

Fig. 2.13  Digital image of interferogram

Box 2 represents the digital pre-processing of the hologram. The hologram data is scanned into computer by means of digital camera in automated systems, or by means of scanner in smaller scale experimental set-up. The digital image is filtered using noise rejection and intensity balancing algorithms [6].

Numerical identification of interference bands in the pre-processed digital image is presented in Box 3 (Fig. 2.13). The output of this step is used as a direct input for the mathematical model processing algorithm (Box 5).

As the formation of interference bands is highly dependent on the geometrical parameters of the experimental set-up, these parameters are a priory calculated

**Fig. 2.14** Deformation
vector of the vibrating plate

before the experiment and feeded into mathematical model algorithm (Box 5) together with the data of interference bands [7, 8]. That's why the developed interferometry data identification system is especially suitable for automated control system—once the geometrical parameters are defined for a certain type of object, the further analysis of a series of objects may be fully automatised as seen from Fig. 2.11.

When the mathematical model is formed, numerical finding vector of unknowns is performed (Box 6) [9]. Further the quantitative parameters of the object's change may be presented in the graphical format (Box 7)—Fig. 2.14.

### 2.1.4  The Mathematical Model of the Optical Measurement

The geometry of used vibration measurement scheme is presented in Fig. 2.15, where $i$ some point with the axes $r$, $t$, $z$ of the orthogonal system shown.

$R$ is the vector of spatial vibrations of the $i$-th point of the pjezo transformer; $U$, $V$, $W$ are the components of the vector of spatial vibrations of the $i$-th point in the directions of the coordinate axis $r$, $t$, $z$, respectively; $l$ is the unit vector of lightening of point $i$; $m$ is the unit vector of observation of point $i$; $\alpha$, $\beta$ are the angles of unit vectors of lightening and observation with the coordinate axis $r$, respectively; $\gamma$, $\theta$ are the angles between the coordinate axis $z$ and the unit vectors of lightening and observation, respectively.

We consider that the spatial vibrations of the surface point $i$ of the analysed body are described as:

$$\bar{R}_i(\tau) = U_i(\tau)\,\hat{i} + V_i(\tau)\,\hat{j} + W_i(\tau)\,\hat{k} \qquad (2.10)$$

where $\tau$ is time.

The tangential $U$, $V$ and normal $W$ components of the vector $\bar{R}_i(\tau)$ at the point $i$ are expressed as follows

$$U(\tau) = U_0^i \cos(\omega\tau + \alpha_i), \quad V(\tau) = V_0^i \cos(\omega\tau + \beta_i),$$
$$W(\tau) = W_0^i \cos(\omega\tau + \gamma_i) \qquad (2.11)$$

**Fig. 2.15** The scheme of optical measurement

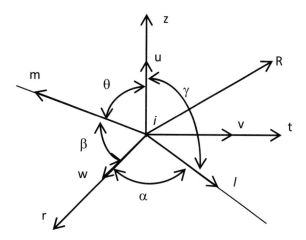

where $U_0^i, V_0^i, W_0^i$ are the amplitudes of forced vibrations at point $i$ in coordinates $r$, $t$, $z$, respectively.

The amplitudes of forced vibrations are expressed by representing them through the eigenmodes of vibrations [10]:

$$U_0^i = \sum_{j=1}^{k} A_j^u F_{ij}^u;$$

$$V_0^i = \sum_{j=1}^{k} A_j^v F_{ij}^v; \qquad (2.12)$$

$$W_0^i = \sum_{j=1}^{k} A_j^w F_{ij}^w;$$

where $F_{ij}$ is the amplitude value of the $j$-th eigenmode of vibration at point $i$, which is calculated according to the analytical expressions by taking into account the conditions of fastening of the analyzed body, $A_j$ is the influence coefficient of the $j$-th eigenmode of vibrations, $k$—the number of given eigenmodes of vibrations; of course, by taking into account the presented relationships it is clear that in order to calculate the components of the vector of spatial vibrations it is necessary to determine $F_{ij}^u, F_{ij}^v, F_{ij}^w, A_j^u, A_j^v, A_j^w, \alpha_i, \beta_i, \gamma_i$.

The values $F_{ij}^u, F_{ij}^v, F_{ij}^w$ are calculated according to the known analytical expressions that are used in theory of vibrations for the calculation of the amplitudes of vibrations of eigenmodes by taking into account the geometry of the body analyzed and the boundary conditions of its fastening [9].

The parameters $A_j^u, A_j^v, A_j^w, \alpha_i, \beta_i, \gamma_i$ are to be determined from the measured experimental data.

According to the characteristic function of distribution of the interferention bands on the surface the following nonlinear algebraic equation is constructed:

$$\frac{\Omega^i \lambda^2}{4\pi} = \begin{bmatrix} \left(\sum_{j=1}^{k} A_j^w F_{ij}^w\right)\cos\gamma_i K_r^i \\ + \left(\sum_{j=1}^{k} A_j^v F_{ij}^v\right)\cos\beta_i K_t^i \\ + \left(\sum_{j=1}^{k} A_j^u F_{ij}^u\right)\cos\alpha_i K_z^i \end{bmatrix}^2 + \begin{bmatrix} \left(\sum_{j=1}^{k} A_j^w F_{ij}^w\right)\sin\gamma_i K_r^i \\ + \left(\sum_{j=1}^{k} A_j^v F_{ij}^v\right)\sin\beta_i K_t^i \\ + \left(\sum_{j=1}^{k} A_j^u F_{ij}^u\right)\sin\alpha_i K_z^i \end{bmatrix}^2 \qquad (2.13)$$

where $\lambda$ is wavelength of the laser lighting; $\Omega$ is calculated from the holographic interferograms at the centers of dark interferentional bands; $K_r^i, K_t^i, K_z^i$ are the projections of the sensitivity vector that are calculated by taking into account the optical scheme of the holographic measurement.

The nonlinear algebraic Eq. (2.13) is solved according to the method presented in [10]. Thus, the following equation is derived from time average holographic interferogram data:

$$f_i = \begin{bmatrix} \left(\sum_{j=1}^{k} A_j^w F_{ij}^w\right)\cos\gamma_i K_r^i \\ + \left(\sum_{j=1}^{k} A_j^v F_{ij}^v\right)\cos\beta_i K_t^i \\ + \left(\sum_{j=1}^{k} A_j^u F_{ij}^{vu}\right)\cos\alpha_i K_z^i \end{bmatrix}^2 - \begin{bmatrix} \left(\sum_{j=1}^{k} A_j^w F_{ij}^w\right)\sin\gamma_i K_r^i \\ + \left(\sum_{j=1}^{k} A_j^v F_{ij}^v\right)\sin\beta_i K_t^i \\ + \left(\sum_{j=1}^{k} A_j^u F_{ij}^u\right)\sin\alpha_i K_z^i \end{bmatrix}^2 - \left[\frac{\Omega_1}{n}\right]^2 \quad (2.14)$$

We differentiate the obtained equation with respect to the unknowns and construct a matrix the columns of which will be respectively

$$G_j^{(i)} = \frac{\partial f_i}{\partial A_j^u}; \quad G_{j+k}^{(i)} = \frac{\partial f_i}{\partial A_j^v}; \quad G_{j+2k}^{(i)} = \frac{\partial f_i}{\partial A_j^w}; \quad j = 1, 2, \ldots, k; \qquad (2.15)$$

$$G_{1+3k}^{(i)} = \frac{\partial f_i}{\partial \alpha_i}; \quad G_{2+3k}^{(i)} = \frac{\partial f_i}{\partial \beta_i}; \quad G_{3+3k}^{(i)} = \frac{\partial f_i}{\partial \gamma_i};$$

If the number of holographic interferograms is made for the point $i$ for different angles of lightening is $q$ and the total number of the data sets is formed, then the dimensions of the matrix $G$ will be $q \times 3(2k + 3)$ [10].

For the given vector of unknowns

$$B = (A_1^u, A_2^u, \ldots, A_k^u, A_1^v, A_2^v, \ldots, A_k^v, A_1^w, A_2^w, \ldots, A_k^w, \alpha, \beta, \gamma) \qquad (2.16)$$

we will seek for the solution of nonlinear algebraic equation by using iterations in the following form:

$$\sigma = \Gamma^{-1} P$$

$$\Gamma_{1j} = \sum_{i=1}^{q} G_1^{(i)} G_j^{(i)}; \quad P_j = -\sum_{i=1}^{q} f_i G_j^{(i)}, \quad j = 1, 2, \ldots (3 + 3k) \qquad (2.17)$$

Numerical solution of Eq. (2.17) produces a spatial deformation vector, which may be represented in a graphical format. The solution of the system enables the reconstruction of special surface deformations of the measured object.

### 2.1.5   Vibration-Assisted Spring-Loaded Micro Spray System. Design and Principle of Operation

The spring-loaded microspray system for drug supplying in vessels is presented. The spring-loaded microspray system consist from the rigid steel spring made of turns without gaps and capable to ensure the system tightness in case of drug supplied under fixed pressure to the sealed spring. The spring-loaded batcher is shown in Fig. 2.16.

Let us suppose that the inlet opening of the spring-loaded microspray system is at the middle of the spring. Another end of the spring *1* is tightened and fixed to transverse vibration vibrator *4*.

When the spring is at rest it does not leak out the liquid drug between the turns (the close contact between the turns provides the tightness of the spring).

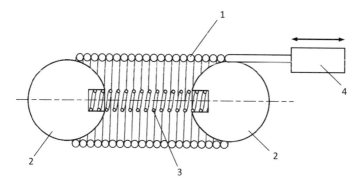

**Fig. 2.16** Spring-loaded microspray system: *1* spring; *2* ball; *3* connecting spring; *4* transverse vibration vibrator

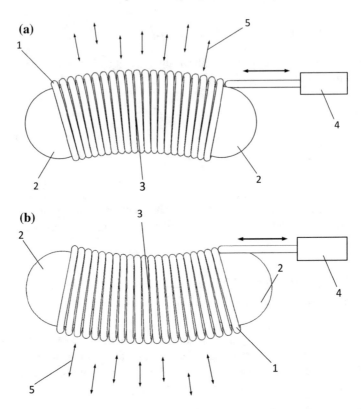

**Fig. 2.17** Spring loaded micro spray system: **a** micro spray to spring moving *upper* direction; **b** micro spray at spring is moved to *down* position

Then, when the transverse vibrations are excited by the help of the vibrator in the form of standing wave in the spring, the spaces between the turns appear, which provide the possibilities for the drug leak out. The half-wave are excited in order to ensure that their amplitude peak phases appeared at the liquid centers of inlet manifold (Fig. 2.17).

It is obvious, that when the piston moves down the rarefaction is caused, which intakes the liquid drug into the vessels.

This is shown only one of the possibilities to arrange the spring loaded micro spray system. The other solution could be to arrange the spring-loaded system in case when we need to excite transverse vibrations, e.g. in the shape of a single half-wave. This would provide the possibility for a separate spring micro spray system loaded to operate independently.

## 2.1.6 *Theoretical Substantiation of Possibilities for the Batcher Functioning*

The rigid coiled spring could be considered as a duct. Let us suppose, that within the range of spring strains analyzed, the material elasticity is constant, therefore, dependence on the strain amount from the applied force is directly proportional. If the spring is affected by the axis strength force, the existing winding area will be proportional to the spring elongation.

The increased surface of elongated spring will be determined, when it is coiled into the arc. The calculation scheme is presented in Fig. 2.18. In the inner part of bended spring the turns touch each other tightly. Hc I the inner arc curvature range is $\rho_0$. It is equal to:

$$\rho_0 = \frac{L}{\pi}. \tag{2.18}$$

The length of the arc L is equal to:

$$L = \frac{2\pi\rho_0}{2} = \pi\rho_0. \tag{2.19}$$

The outer part of the arc between the turns will have the gap $\delta$, which being in the shape of spiral, decreases to 0 in the inner part of the arc. Thus, the gap of spiral shifting width gap is produced.

**Fig. 2.18** Calculation scheme of the spring elongation

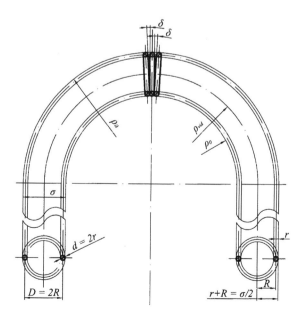

The outer arc radius $\rho_{ext}$ is equal:

$$\rho_{ext} = \rho_0 + 2(R+r). \tag{2.20}$$

The outer arc length $L_{ext}$ is:

$$L_{ext} = \pi\rho_{ext} = \pi[\rho_0 + 2(R+r)] = \pi(\rho_0 + \sigma), \tag{2.21}$$

where, $\sigma$—spring-duct diameter. It is equal to:

$$\sigma = 2(R+r). \tag{2.22}$$

Thus outer arc length $L_{ext}$ is:

$$L_{ext} = \pi\left(\frac{L}{\pi} + \sigma\right) = L + \pi\sigma. \tag{2.23}$$

Outer arc elongation $\Delta L$ is equal to:

$$\Delta L = L_{ext} - L = L + \pi\sigma - L = \pi\sigma. \tag{2.24}$$

Thus average elongation of the spring $\Delta L_{ave}$ is equal to:

$$\Delta L_{ave} = \frac{\Delta L}{2} = \pi(8+r). \tag{2.25}$$

Increased surface of average elongated spring $\Delta S_{ave}$ will be equal to:

$$\Delta S_{ave} = 2\pi R\Delta L_{ave} = 2\pi^2 R(R+r). \tag{2.26}$$

This is the space for the leak out of the part of fuel.

Let us analyze the case, when the spring-duct axis is in the shape of curve, which is presented in Fig. 2.19.

**Fig. 2.19** Spring axis as curve

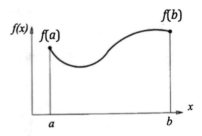

In general case between $f(a)$ and $f(b)$:

$$l = \int_a^b \sqrt{1 + [f'(x)]^2}. \qquad (2.27)$$

If the excited vibrations are in the shape of sine, the half of its length $L_p$ will be (Fig. 2.19):

$$L_p = \int_0^{1/4} \sqrt{1 + [\sin' x]^2} = \int_0^{1/4} \sqrt{1 + \cos^2 x}. \qquad (2.28)$$

The spring-duct elongation half-waves $\Delta L_p$ will be:

$$\Delta L_p = L_p - \frac{l}{4}. \qquad (2.29)$$

This elongation of the spring affects the increase of its inner surface:

$$\Delta S_{ave} = 2\pi R \Delta L. \qquad (2.30)$$

Thus, the outer surface S area change could be expressed as:

$$S = A_0 \cos\left(\frac{2\pi}{\lambda} x\right) \sin(2\omega t), \qquad (2.31)$$

$$\omega = 2\pi f, \qquad (2.32)$$

where: $A_0$—maximum amplitude of standing waves; $x$—spring-duct coordinate along axis; $\lambda$—length of wave; $f$—frequency, Hz.

Thus, we could confirm, that the higher the amplitude of spring vibration, the wider the space between the spring turns and more fuel will leak out between them.

### 2.1.7 Experimental Analysis of the Spring

In order to calculate amplitude of vibrating spring the methodology is presented in papers [7, 9, 10].

In Fig. 2.20 it is shown optical scheme for recording holographic interferograms of the vibrating spring: *1*—vibrating spring; *2*—high-frequency signal generator; *3*—amplifier. The signal monitoring means are; *4*—frequency meter, *5*—the voltage amplitude of the power supply is monitored by the voltmeter. The optical scheme includes a holographic table with a helium-neon laser which serves as a

**Fig. 2.20** Optical scheme of
the laser holographic
interferometry system:
*1* tubular working tube,
*2* high-frequency signal
generator, *3* amplifier,
*4* frequency meter,
*5* voltmeter, *6* laser,
*7* beam splitter, *8, 9* mirror,
*10, 11* lens, *12* photographic
plate, *13* recorder

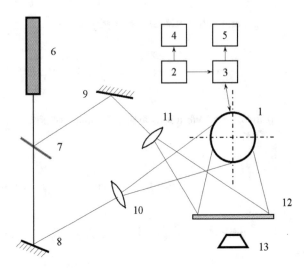

source of coherent radiation. At first the beam from the optical laser *6* splits into two
coherent beams and one of them is passing through the beam splitter *7*. The another
one, so called object beam, reflected from the mirror *8*, and widespread lens *10* and
illuminates the surface of the vibrating spring *1* and, after reflecting from it, illuminates the photographic plate *12*. The reference beam, reflected by the mirror *9*,
and by the lens *11*, illuminates the holographic plate *12* where the interference of
these two beams is recorded.

The characteristic function defining distribution interference on the surface of the
vibrating spring is presented in (2.33).

$$M_T = \lim_{T \to \infty} \frac{1}{T} \int_0^T \exp\left(i\left(\frac{4\pi}{\lambda}\right)Z(x)\sin \omega t\right)dt = J_0\left(\left(\frac{4\pi}{\lambda}\right)Z(x)\right) \qquad (2.33)$$

where *T*—the exposure time vibrating spring onto the hologram, $(T \gg 1/\omega)$; $\omega$—
the frequency of vibration of spring, $\lambda$—the laser wavelength of used for recording
holographic interferogram; $J_0$—zero order Bessel function of the first type.

Then, the resulting intensity *I* of the point $(x, y)$ on the holographic interferogram
of vibrating spring is follows:

$$I(x,y) = a^2(x,y)|M_T|^2, \qquad (2.34)$$

where $a(x, y)$ defines the distribution of the amplitude of the incident laser beam.
The usage of the method of time averaging holographic interferometry allows to
measure steady state vibration. Results of experimental analysis vibrating spring are
presented in Fig. 2.21.

**Fig. 2.21** Results of experimental analysis: holographic interferogram of vibrating spring at frequency 1.24 kHz **a**, distribution amplitude of vibration of spring **b**

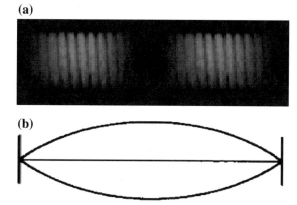

(a)

(b)

## 2.2 Numerical–Experimental Method for Evaluation of Geometrical Parameters of Periodical Microstructure

Optical modulator is a device that modulates or varies the amplitude of an optical signal in a controlled manner. Optical modulator generates desired intensity, color and the like in the passing light by changing optical parameters such as the transmission factor, refractive index, reflection factor, degree of deflection and coherency of light in the optical system according to the modulating signal. A constituent part of the modulator is a diffractive optical element (DOE).

Different methods and materials are used to produce diffraction gratings for DOEs. On the other hand two-dimensional or three-dimensional periodic structures of micrometer-scale period are widely used in microsystems or their components, e.g. as elements of micro-fluidic devices. Long deep groves (in optical terms—phase diffraction grating) can be used as elements for microscale synthesis, processing, and analysis of chemical and biological samples that require manipulation of microscopic volumes of liquids. Usually this can be accomplished with chips with micro-channels and microreactors.

Replication technologies such as embossing, molding and casting are highly attractive for the fabrication of surface relief holograms and diffractive optical elements microstructures [11]. The combination of replication technology with other processes such as dry etching and thin film coating can offer new possibilities in the mass production. The major replication technologies that are in use today [hot embossing, injection molding and casting (or UV embossing)] employ different types of polymers. Polymers are receiving global attention for a myriad of planar photonic and optoelectronic applications including optical interconnects [12], switches [13], splitters [14], and surface relief structures [15]. This is a direct result of the relative ease and cost effectiveness with which planar polymeric structures can be fabricated, with respect to semiconductor and oxide analogs, while maintaining the requisite performance levels.

For characterization of geometrical parameters of such microrelief structures usually various construction microscopes are used, mostly scanning electron or probe microscopes [16]. These direct methods are sometimes destructive and hardly can be employed for in situ analysis. Therefore indirect optical interference or diffraction methods are used widely [17].

Measuring diffraction efficiency for the visible light is known as an indirect method to evaluate geometrical parameters of diffraction gratings [18, 19]. Diffraction efficiency is one of the crucial properties of the optically variable devices such as kinegrams or 3D holograms [20] that are used widely during last decade to provide document security [20]. High efficiency diffraction grating is important as well in a variety of applications, such as optical telecommunications, lithography, and laboratory spectroscopy [21].

From this point of view optical methods are very flexible and efficient in control where dimensions of periodic structures are in micrometer range.

## 2.2.1 Concept of Indirect Method for Evaluation of Geometrical Parameters of Periodical Microstructure

All periodical microstructures formed in optical materials are characterized by relative diffraction efficiencies. Relative diffraction efficiency $RE_{i,j}$ is defined as ratio of intensity of diffracted light $I_{i,j}$ to the i-th diffraction maxima and j-th illumination angle with intensity $I_j$ of reflected light or transmitted through specimen without micro relief to j-th illumination angle:

$$RE = \frac{I_{i,j}}{I_j}. \tag{2.35}$$

Comparison of modeled diffraction efficiencies with experimental results could be used to control variation of geometrical parameters of periodical microstructure during technological process. Difference $c_{SE}$ (Fig. 2.22) between numerical and experimental results is calculated using least squares method:

**Fig. 2.22** Difference c versus depth of periodical microstructure (sinusoidal profile period d = 4 μm) for green laser (λ = 532 nm)

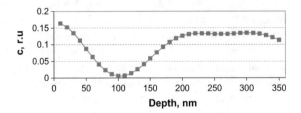

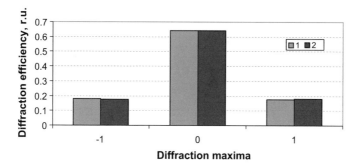

**Fig. 2.23** Experimental (1) and theoretical (2) relative diffraction efficiencies of periodical microstructure (sinusoidal profile, depth 105 nm, period d = 4 μm) for the green laser (λ = 532 nm)

$$c = \frac{1}{k \cdot n} \sum_{j=1}^{k} \sum_{i=1}^{n} \left( RE_{i,j}^{T} - RE_{i,j}^{E} \right)^2; \qquad (2.36)$$

$RE_{i,j}^{T}, RE_{i,j}^{E}$—numerical and experimental relative diffraction efficiencies to the i-th diffraction maxima and j-th illumination angle.

Depth of periodical microstructure is in the minimum point of curve c, where the difference between experimental and numerical results is the smallest. Application of this method enables evaluation of geometrical parameters with an error of less then 5%. Good fit of experimental and theoretical relative diffraction efficiencies of periodical microstructure (sinusoidal profile, depth 105 nm, period d = 4 μm) for green laser (λ = 532 nm) is illustrated in Fig. 2.23. Calculations were confirmed with atomic force microscope NANOTOP 206.

This method could be used for the non-destructive control of variation of geometrical parameters of periodical microstructure during technological process.

## 2.2.2 Evaluation of Geometrical and Optical Parameters of Periodical Microstructure

Figure 2.24 presents AFM photographs of a matrix in silicon (a), matrix polymer replica (b), matrix in nickel stamp (c) and matrix hot embossed in Al metalized PMMA layer on PET (d). Table 2.1 summarises the main parameters [period (d), depth (h) and modulation coefficient (μ)] of the investigated diffraction gratings. The columns denoted as "Measured by AFM" present experimental values of the grating. As one can see modulation of the diffraction grating is lost during the replication steps.

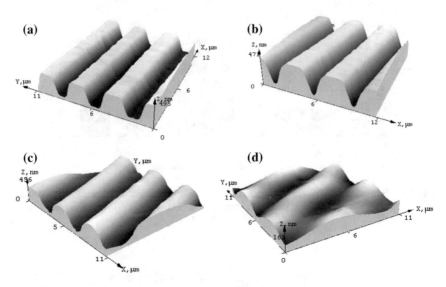

**Fig. 2.24** *Top-left* AFM photographs of matrix in silicon. *Top-right* matrix polymer replica. *Bottom-left* matrix in nickel stamp. *Bottom-right* matrix hot embossed in Al metalized PMMA layer on PET

**Table 2.1** Main parameters of the investigated diffraction gratings

| Diffraction grating | Period d, μm | | Depth h, μm | | Modulation, $\frac{h}{d}$ | Groove profile |
|---|---|---|---|---|---|---|
| | Measured by AFM | Simulated | Measured by AFM | Simulated | | Measured by AFM |
| Si master-matrix | 4 | 4.1 | 0.5 | 0.535 | 0.125 | Trapezoidal |
| UV replica | 4.2 | 4.1 | 0.47 | – | 0.112 | Trapezoidal |
| Ni stamp | 4.2 | 4.1 | 0.4 | – | 0.095 | Sinusoidal |
| Replica T = 120 °C | 4.4 | 4.3 | 0.2 | 0.195 | 0.045 | Sinusoidal |

UV curing plays an important role in cementing of optics or for fabrication of replicated optics [21]. For replicated optics, shrinkage and form modifications are usually measured by comparing the replica to the mould geometry after the curing process has finished [17]. One can see that, producing of UV replica is related to the production of the reverse grating of the silicon matrix with the higher value of the period. This fact is known [22, 23] and is related to the expansion of UV replica. According to [23] in the case of big patterns, a large amount of resist has to be displaced over a large distance. Thus the polymer in the middle of these patterns does not flow but it is compressed and stores stresses. This compressed polymer reacts elastically and when the force is removed, it recovers.

The biggest looses of the modulation are found during the thermal embossing process. It should be noted as well that originally trapezoidal diffraction grating during production of the Ni stamp and hot embossing is transformed to the sinusoidal one.

To follow variations of the geometrical parameters, the diffraction efficiencies of the periodic structures were registered after the main technological steps. Figure 2.25 (curves 1, 2) illustrates experimental dependence of the relative efficiency of the first diffraction maximum $(RE_1)$ and all maxima $(RE_S)$ versus angle of incidence for the silicon master matrix. Angles of incidence were varied between 2° and 47°. Curve 3 (Fig. 2.25) illustrates variation of the relative efficiency as simulated with the PCGrate-SX6.0 programme. During these calculations trapezoidal diffraction grating in silicon was considered as rectangular one and the ridge width as well as depth of the grating was varied systematically to fit the experimental curve (2) in Fig. 2.25. Geometrical parameters of the simulated grating are presented in Table 2.2 where they are compared with the corresponding parameters of the silicon master matrix measured by AFM. Such an approach allows to calculate angular dependence of the relative diffraction efficiency with high determination coefficient $(R^2 = 0.99)$ and to reconstruct the geometrical parameters of the diffraction grating from the optical measurements with accuracy better than 10%. In such evaluations diffraction efficiency of the first diffraction maximum is informative enough and higher maxima may be ignored during the consideration.

Changes in the shape of periodic structure during the replication can be easily detected by measuring angular dependence of the diffraction efficiency as it is shown in Fig. 2.26. In this case angular dependences of diffraction efficiencies of the structure (presented in Fig. 2.24) are depicted (curve (1) corresponds to the Si master matrix, curve (2)—matrix in the stamp, curve (3)—matrix embossed in

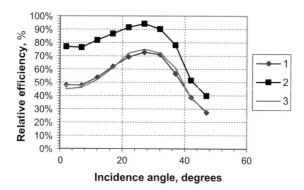

**Fig. 2.25** Relative efficiencies of the diffraction grating in silicon (master matrix) (d = 4 µm, h = 0.5 µm, a = 1.9 µm) versus angle of incidence: (1) experimentally measured for the first diffraction maximum, (2) experimentally measured for the six diffraction maxima, and (3) computer simulated for the first diffraction maximum

**Table 2.2** Comparison of the simulated geometrical parameters of the silicon master matrix with the corresponding parameters measured by AFM

| Material | Results | Period d, μm | Ridge width, a | | Side wall deviation angle, φ | Depth h, μm |
|---|---|---|---|---|---|---|
| | | | μm | r.u.[a] | | |
| Crystalline Si(100) | Measured by AFM | 4 | 1.9 | 0.48 | 82° | 0.5 |
| | Simulation | 4.1 | 2.1 | 0.51 | 90° | 0.535 |

[a]Ridge width in r.u. is calculated as $\frac{a}{d}$

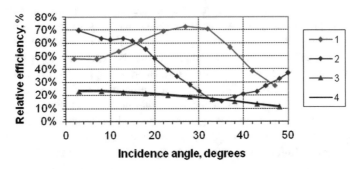

**Fig. 2.26** Angular dependencies of the relative efficiencies of the diffraction grating measured at different technological steps: 1—Si master matrix, 2—nickel matrix, 3—matrix hot embossed in Al metalized PMMA layer on PET, 4—simulated diffraction grating

Al/PMMA structure). Curve (4) presents calculated angular dependence of the grating produced in Al/PMMA structure using the experimentally defined (by AFM) parameters of the grating. One can see that low modulation values (as it is defined from the AFM measurements, Table 2.1) bring to the low value of diffraction efficiency. High value of determination coefficient (0.95) for the experimental curve (3) and simulated curve (4) illustrate that sinusoidal profile describes well the real profile of the grating. A column "Simulated" in Table 2.1 summarizes main results of the computer simulation that can be compared with the AFM measurement results.

Such an approach appears as an efficient method to analyze hot embossing process. Figure 2.27 presents the dependence of the relative diffraction efficiency versus embossing temperature at constant angle of incidence of analyzing light. Figure 2.28 illustrates angular dependencies of the relative efficiency for two temperatures of embossing. In all investigated cases [different periods of diffraction grating curves (1, 2) and curves (3, 4) and different temperatures of embossing curves (1, 3) and (2, 4)] relative efficiency of resultant replica in Al/PMMA are well described by sinusoidal profile (as it was shown in Fig. 2.26). One can see (Fig. 2.27) that sinusoidal profile from Ni matrix is transferred efficiently to the

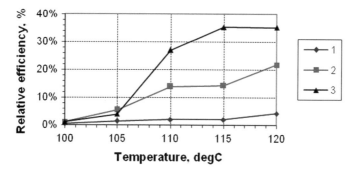

**Fig. 2.27** Dependence of the relative diffraction efficiency versus embossing temperature for the gratings of h = 0.5 μm and different period: 1—d = 2 μm, 2—d = 4 μm, 3—d = 5.6 μm (angle of incidence of light 3°)

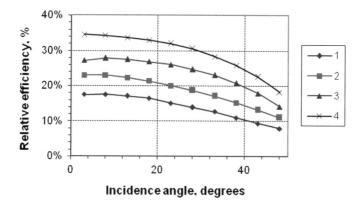

**Fig. 2.28** Angular dependencies of relative efficiencies of gratings of h = 0.5 μm and different period: 1, 2—d = 4 μm, 3, 4—d = 5.6 μm for two embossing temperatures 1, 3—T = 110 °C, 2, 4—T = 120 °C

PMMA within the interval of embossing temperatures 110–120 °C. Within this interval of temperatures relative efficiency reaches maximum value within the technologically compatible processes (p = 0.3 MPa, t = 2 s). One can see as well that this process is sensitive to the modulation ratio of the grating. Our measurements reveal well documented fact [24] that in lower temperature range elastic response of the polymer dominates and in the region close to the polymer glass transition temperature plastic flow of the polymer contributes to the efficient shape transfer. According to our optical measurement results keeping constant pressure and time of embossing, rheological properties of polymer are expressed better for the higher period grating. The best relative efficiency was found for the diffraction grating of 5.6 μm period embossed at 120 °C temperature.

## 2.2.3  Evaluation of Geometrical Parameters with High Aspect Ratio

However, when the depth (h) of periodic microstructure is higher than the order of coherent radiation wavelength ($\lambda$) used for investigation of depth, the periodical repeatability of theoretical results is obtained (2.29). Therefore there are several sets of geometric parameters with the good agreement to experimental. In this case, without the usage of additional measuring instruments such as atomic force microscopy or scanning electron microscope, it is not possible clearly to identify the depth of the periodic microstructure. Therefore the methodology, based on comparison of optical and numerical results, for determination of deep (micrometers queue depth) periodic microstructures depth was developed

Analyzing the $RE_{i,j}$ dependence on the depth (Fig. 2.29), in the region of the small depth one can see clearly influence of both factors (angle of incidence and wavelength). One can see that applying higher wavelength results in the shift of these dependencies to the right and this shift is more pronounced for the higher angle of incidence.

According to the simulation results (Fig. 2.29), dependence of the relative diffraction efficiency of the zero diffraction maximum versus depth of grating is a periodical function. This phenomenon can be easily understood in terms of optical path length difference produced in the reflection diffraction grating (Fig. 2.30). One can understand that additional optical path length difference distance $\Delta$ in the phase diffraction grating can be calculated using a geometrical model. This optical path

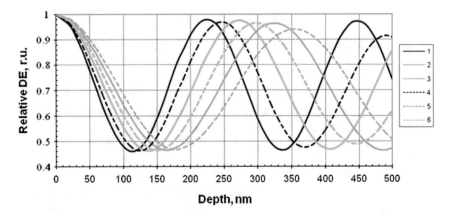

**Fig. 2.29** Simulated dependence of the relative diffraction efficiency of the zero order diffraction maxima on the depth of groove of the phase diffraction grating (diffraction grating in silicon, d = 12 μm) for different light wavelength (441.6, 532, 632.8 nm) and incidence angles (5°, 25°): 1 —wavelength 441.6 nm and illumination angle 5°; 2—wavelength 532 nm and illumination angle 5°; 3—wavelength 632.8 nm and illumination angle 5°; 4—wavelength 441.6 nm and illumination angle 25°; 5—wavelength 532 nm and illumination angle 25°; 6—wavelength 632.8 nm and illumination angle 25°

**Fig. 2.30** Geometrical model explaining optical path length difference in a rectangular phase reflection diffraction grating

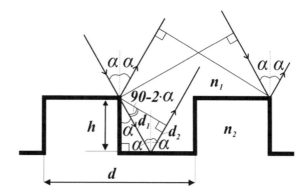

length difference depends on the incidence angle $\alpha$, refraction index of medium ($n_1$) and depth (h) of the phase diffraction grating:

$$\Delta = n_1(d_1 + d_2);\tag{2.37}$$

$$\Delta = \frac{n_1 h}{\cos \alpha}(1 + \sin(90 - 2\alpha)).\tag{2.38}$$

One can understand as well that constructive interference condition:

$$\Delta = m\lambda;\quad m = 0, 1, 2, \ldots\tag{2.39}$$

will define depth of the diffraction grating (for fixed angle of incidence) where maximum value for the zero maximum will take place. Changing the angle of incidence ($\alpha$) or wavelength will result in the change of $\Delta$, and, as a result, change of position of maximum versus depth of the grooves. These dependencies (integer number of wavelengths versus depth of the grating) are summarized in Fig. 2.31. The cross of the horizontal line (Fig. 2.31) corresponding $\Delta = 1\lambda$ or $\Delta = 2\lambda$ in the case of deep grooves allows predicting depth of the grating corresponding to the maximum intensity of the zero maximum (i.e. in this case for 441.6 nm wavelength and $5°$ illumination angle they correspond to h = 0, 220 and 440 nm). These evaluations are in good correlation with the simulation results done with "PCGrate" (Fig. 2.29). One can understand that grating of the different depth then discussed ones will produce maxima at different angle of incidence.

In similar way, depths of diffraction grating providing minimum value of the zero maximum can be easily calculated putting

$$\Delta = (2m+1)\lambda/2;\quad m = 0, 1, 2, \ldots\tag{2.40}$$

This condition defines depth of diffraction grating corresponding to the minimum of the zero order maximum (or maximum of the first order maximum). By the way, this condition defines optimal depth of the diffraction grating working as a beam splitter (zero order maximum is equal to zero and the first order maxima have

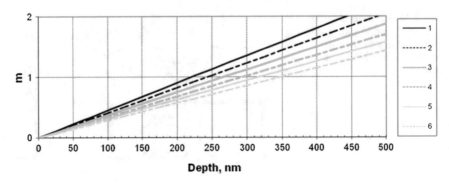

**Fig. 2.31** Number of integer wavelengths in optical path length difference versus depth of the rectangular phase reflection diffraction gratings produced in silicon (calculated for angles of incidence 5°, 25° and wavelengths 441.6, 532, 632.8 nm): 1—wavelength 441.6 nm and illumination angle 5°; 2—wavelength 441.6 nm and illumination angle 25°; 3—wavelength 532 nm and illumination angle 5°; 4—wavelength 532 nm and illumination angle 25°; 5—wavelength 632.8 nm and illumination angle 5°; 6—wavelength 632.8 nm and illumination angle 25°

maximal value). This condition in our case looking for the first optimal situation (Fig. 2.29) should be satisfied at the depth equal to 110 nm.

Measuring relative diffraction efficiencies with three wavelengths one can get additional information—relative diffraction efficiency dependence on the wavelength. For example, comparing differences between relative diffraction efficiencies measured with red green and blue light wavelengths we have three additional parameters—differences between relative diffraction efficiencies: for red–green, red–blue, green–blue light. This situation is explained in Fig. 2.32. Combining values of experimental and theoretical relative diffraction efficiencies as well as experimental and theoretical differences between relative diffraction efficiencies measured with different wavelengths one can avoid problems related to the periodical dependence of $RE_{i,j}$ versus depth. In such case three more parameters related to the different wavelength relative diffraction efficiency should be analyzed:

$$RD_{R-G,j} = \frac{RE_{R,j} - RE_{G,j}}{RE_{R,j}}, \qquad (2.41)$$

$$RD_{R-B,j} = \frac{RE_{R,j} - RE_{B,j}}{RE_{R,j}}, \qquad (2.42)$$

$$RD_{G-B,j} = \frac{RE_{G,j} - RE_{B,j}}{RE_{G,j}}, \qquad (2.43)$$

where:

$RD_{R-G,j}$                    is relative difference between relative diffraction efficiencies of zero maxima for wavelengths 632.8 and 532 nm;

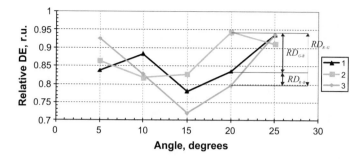

**Fig. 2.32** Dependence of the experimental relative diffraction efficiencies of zero order maxima measured with three lasers (632.8, 532, 441.6 nm) versus illumination angle: 1—wavelength 441.6 nm, 2—wavelength 532 nm; 3—wavelength 632.8 nm

| $RD_{R-B,j}$ | is relative difference between relative diffraction efficiencies of zero maxima for wavelengths 632.8 and 441.6 nm; |
| $RD_{G-B,j}$ | is relative difference between relative diffraction efficiencies of zero maxima for wavelengths 532 and 441.6 nm; |
| $RE_{R,j}, RE_{G,j}$ and $RE_{B,j}$ | are relative diffraction efficiencies of zero maxima for red, green and blue lasers. |

## 2.2.4  Investigation of Microstructures of High Aspect Ratio

Validation of the proposed model was performed on a system of channels of the micro-fluidic device (reflecting diffraction grating). SEM view of the fabricated micro-fluidic device in silicon is demonstrated in Fig. 2.33. According to the SEM measurements, the period of microchannels was 12 μm, ridge width 9 μm and groove width 3 μm.

The modeling results of reflection diffraction grating were used for indirect depth evaluation of the deep periodic structures—micro-fluidic device in silicon. Two approaches were used for the depth determination of microstructures:

I—comparing values of theoretical ant experimental relative diffraction efficiencies $RE_{i,j}$ for three different wavelengths ($\lambda \in (632.8, 532, 441,6$ nm)) and five values of incidence angle ($j \in (5°, 10°, 15°, 20°, 25°)$).

II—comparing experimental and theoretical differences between relative diffraction efficiencies $\left(RD_{R-G,j}, RD_{R-B,j}, RD_{G-B,j}\right)$ obtained with three different wavelengths for five values of incidence angle ($j \in (5°, 10°, 15°, 20°, 25°)$).

In the first case experimental values of angular dependence of relative diffraction efficiency of the second, first and zero order maxima were compared with the computer simulated ones for different depth of diffraction gratings for various illumination angles of visible light.

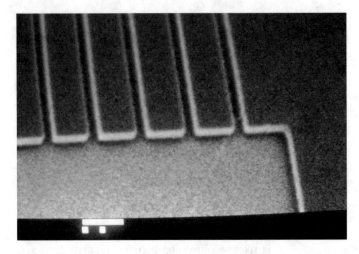

**Fig. 2.33** SEM view of channels of micro-fluidic device (the reservoir connecting the channel system), scale bar 10 μm

During this process we have calculated the coefficients of correspondence to the different depth $c_{depth,j}^{RE}$ of simulated diffraction gratings (for different irradiation angles) to the corresponding experimental values of relative diffraction efficiency (2.44). To find the depth of the grooves these coefficients calculated for different irradiation angles were averaged and average value $\left( a_{depth}^{RE} \right)$ was calculated according to (2.45). The dependence of average of coefficients versus range of possible depth then was plotted (Fig. 2.34).

$$c_{depth,j}^{RE} = \frac{1}{n} \sum_i \left( RE_{i,j}^T - RE_{i,j}^E \right)^2 ;$$
(2.44)

$$a_{depth}^{RE} = \frac{1}{k} \sum_j c_{depth,j}^{RE} ;$$
(2.45)

where $RE_{i,j}^T$ and $RE_{i,j}^E$ are theoretical and experimental relative diffraction efficiencies of i-th maxima at j-th illumination angle, n is number of measured maxima and k is number of illumination angles.

One can see that $c_{depth,j}^{RE} = f(h)$ is a periodical function, i.e. direct comparison of simulation results of relative diffraction efficiency and experimental relative diffraction efficiency in the case of one wavelength enables us to define depth of the grooves with some period only. To avoid this uncertainty multiple wavelengths were used in this case. Comparing experimental values of relative diffraction efficiency of 5 diffraction maxima (±2, ±1, 0) for three different wavelengths and corresponding theoretical relative diffraction efficiencies, the depth of the channels

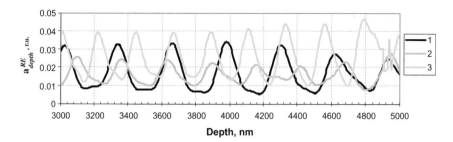

**Fig. 2.34** Average sum of squared differences between simulated and experimentally measured relative diffraction efficiencies for different wavelengths lasers versus depth of the rectangular phase reflection diffraction gratings produced in silicon: 1–632.8; 2–532; 3–441.6 nm

was defined as a minimum of the average of $a_{depth}^{RE} = f(h)$ for all three different wavelengths. According to Fig. 2.34 this minimum is in the vicinity of 3790 nm.

Results obtained in the first approach were compared with the calculations made using the second approach. In this case three functions $a_{depth}^{R-G} = f(h), a_{depth}^{R-B} = f(h)$ and $a_{depth}^{G-B} = f(h)$ defined according to Eqs. (2.46)–(2.48) were plotted as function of depth of grooves as it shown in Fig. 2.35.

$$a_{depth}^{R-G} = \frac{1}{k}\sum_{j}\left(RD_{R-G,j}^{T} - RD_{R-G,j}^{E}\right)^2 \tag{2.46}$$

$$a_{depth}^{R-B} = \frac{1}{k}\sum_{j}\left(RD_{R-B,j}^{T} - RD_{R-B,j}^{E}\right)^2 \tag{2.47}$$

$$a_{depth}^{G-B} = \frac{1}{k}\sum_{j}\left(RD_{G-B,j}^{T} - RD_{G-B,j}^{E}\right)^2 \tag{2.48}$$

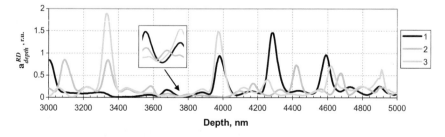

**Fig. 2.35** Average sum of squared differences between simulated and experimentally measured distances of relative diffraction efficiencies for different wavelengths lasers versus depth of the rectangular phase reflection diffraction gratings produced in silicon: $1—a_{depth}^{R-G}$, $2—a_{depth}^{G-B}$, $3—a_{depth}^{R-B}$

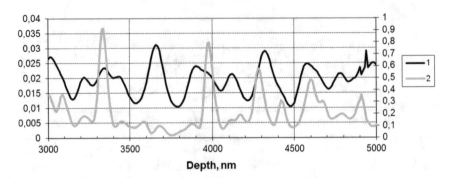

**Fig. 2.36** Average curves of $a_{depth}^{RE}$ (1) and $a_{depth}^{RD}$ (2) versus depth of the rectangular phase reflection diffraction gratings produced in silicon

Comparing experimental values of relative differences between the zero order relative diffraction efficiencies maxima measured with three different wavelengths and corresponding theoretical relative diffraction efficiencies, the depth of the channels was defined as a minimum of all three curves $a_{depth}^{R-G}, a_{depth}^{G-B}, a_{depth}^{R-B}$ (Fig. 2.35) for different wavelengths (632.8, 532 and 441.6 nm) in the vicinity of 3760 nm. The depth of grooves of microstructures can be specified from the minima of average curves $a_{depth}^{DE}$ and $a_{depth}^{RD}$ (Fig. 2.36). It was found that the depth is equal to 3775 nm.

This range of depths (3760–3790 nm) is in good agreement with the results of SEM analysis demonstrating feasibility of the proposed method. Contrary to the direct method (SEM or AFM) the proposed analysis combining computer simulation and experimental results of relative diffraction efficiency versus different angle of incidence and wavelength of employed light can be used for in situ analysis e.g. control of kinetics of microfluidic device.

## 2.3 Polycarbonate as an Elasto-Plastic Material Model for Simulation of the Microstructure Hot Imprint Process

### 2.3.1 Theoretical Background for Finite Element Model of Hot Imprint Process

The aim of the hot imprint model is to help the pattern and mold designer predict the main flat imprint parameters, namely, the loads, temperature and velocity fields, the overall geometric changes of deformed pieces, as well as the optimum conditions for the flat imprint process.

Stoyanov et al. [25] in nano-imprint forming (NIF) process used four steps:

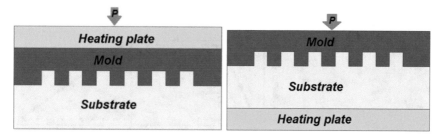

**Fig. 2.37** Schematic imprint process diagram with one heating plate

1. Softening of a thin film of formable material deposited on a substrate;
2. Pressing a rigid mould which has the required (negative) pattern of features onto the formable material;
3. Curing the formable material by cooling down to temperature below glass transition temperature ($T_g$);
4. Releasing the mold.

Depending from tool which is used in experiment, hot embossing process model have one or two heating plates (Figs. 2.37 and 2.38).

Some researches in the modeling and simulation of hot embossing process use three steps. Depending from requirements of the model and expected results, they integrate two steps in one, how heating and imprint, cooling and demolding. It is possible, when created model conform expected and predicted results, and is adequate by experimental date. He et al. [27] in hot imprint process modeling used three steps: molding (include preheating), cooling, demolding.

Typical hot imprint process is shown in Fig. 2.39. Black line represents changes of temperature and dotted line—pressure. In hot embossing process, a thermoplastic polymer is heated over or near its glass transition temperature ($T_g$), and a fine mold is pressed into the polymer. After cooling down below $T_g$ with pressure, the mold is released and the fine pattern on the mold is transferred to the polymer.

**Fig. 2.38** Schematic imprint process diagram with two heating plates

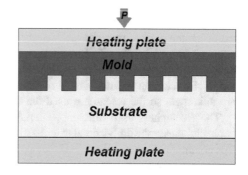

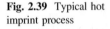

**Fig. 2.39** Typical hot imprint process

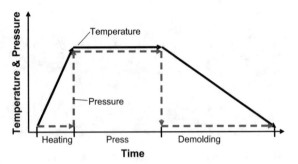

Especially in hot imprint process, cycle time is a very important issue. Substrate temperature is rising depending on the time during the heating process (Fig. 2.39). In this process there is no pressure, it equals 0. At the beginning of the second step, pressure is applied in the model and it stays constant to the end as temperature. At the beginning of the step three, the mold is removed. It means that pressure becomes 0 and temperature of the substrate decreases. Velocity of temperatures decreasing depends from materials and environment.

Some researches simulated and analyzed just one of the hot imprint process steps. Lan et al. [28] analyzed imprint step and filling ratio at different temperature and imprinting time. Song and et al. [29] analyzed only demolding step in thermal imprint lithography. Anyway, the model could be used for detail analysis when it corresponds the experimental results.

Hot imprint process can be performed isothermally and non-isothermally. The isothermal hot imprint means the substrate and mold will be heated to the same temperature during imprint process. The non-isothermal imprints means that the mold and substrate will be heated to different temperature [30]. Isothermal method is the most popular process for the hot embossing of solid polymers. The "solid polymer" refers to the polymer under glassy state and hyperelastic state [31].

Jeong et al. [32] and Juang [26] used isothermal nonlinearity conditions. Juang [26], Yao et al. [33] created nonisothermal hot embossing model. For thick polymer substrates (e.g. 2 mm), nonisothermal embossing is desired to minimize the subsequent cooling time.

Depending on the material properties, imprint temperature, solved problems many mathematical material models are created. Polymers, depending on the temperature can be classified into three states: glass, semi-molten, flow [34]. The temperature range is an answer, which material model will be use considered in the model. Developing a mathematical model, often material properties and parameters are unknown, so the researches used the experimental results.

The researchers used different material models: hyperelastic, viscoelastic, viscoplastic, elastoplastic, visco-elastic-plastic and etc.

Schmid and Carley [35], Day et al. [36] used the hyperelastic model to simulate polymer material properties in the semi-molten state. They indicated that in semi-molten state $T_g < T < T_g + 60$ (where $T_g$ is the glass transition temperature) some amorphous polymer had rubber-like properties.

Krishnaswamy et al. [37] used viscoelastic and viscoplastic models, to simulate the material properties of a ductile crystalline polymer, such as polyethylene (PE), at room temperature.

Lin et al. [38] used nonlinear viscoelastic material model for PMMA, where Young's modulus, thermo-conductivity coefficient and viscosity are depended on temperature, pressure and working time. Nicoli [39] developed a continuum-mechanical constitutive theory aimed to fill the transition state from a visco-elasto-plastic solid-like to a fluid-like response. In Kiew et al. [40] model, the PMMA has elastic-plastic property under 120 °C and viscoelastic property when temperature exceeds 120 °C. This means that PMMA will remain in the solid state when temperature is below 120° and semi-molten state when temperature is above 120 °C. Kim et al. [41] analyzed thermal NIL process using the viscoelastic material model for PMMA with a temperature range $T_g < T < T_g + 40$. Yao et al. [33] used viscoplastic material model for PMMA and solved the coupled flow and heat transfer problem in the nonisothermal embossing process. Jin et al. [42] used combination of two models: non-Newtonian fluid and linear-elastic solid above the glass transition temperature. Hirai et al. [43] applied the Moony-Rivlin model to understand the deformation process of polymer in the nanoimprint process. Young [44], Juang et al. [26] simulated PMMA as viscous fluid model over it glass transition temperature. Yao et al. [33] simulated non-isothermal hot embossing process for PMMA as viscoplastic flow material model. They indicated flow movement and temperature distribution inside the cavity for different cavity thickness. Song et al. [29] used viscoelastic material model for the thermal imprint lithography simulation. Dupaix and Cash [45] created a hyperelastic-viscoplastic constitutive model for amorphous polymer for hot embossing process.

There are a few hot imprint models for polycarbonate (PC). Lan et al. [28] obtained a numerical viscoelastic material model based on generalized Maxwell model for the material near the glass transition temperature ($T_g$) for PC. A number of preliminary tensile stress relaxation tests were carried out at 150 °C at different instantaneous strain to produced stress relaxation curves. These curves were used to estimate model parameters. Kiew et al. [40] converted temperature-stress-strain experimental data points into a matrix expression and got the full scale deformation trend. This method helps the process engineers to gain insights on how the polymer substrate would behave during the hot imprint process. Lin et al. [38] did uniaxial compression experiments at various temperatures and creep test in order to obtain required material parameters for viscoelastic model. The stress/strain data from the compression test was used to specify the short-term elastic properties, i.e. handles the instantaneous displacement behavior of the PMMA during the situation of loading and unloading. The shear stress relaxation modulus from the creep test was used to specify the time-dependent behavior of PMMA.

The hot imprint process can be modeled as a two-dimensional or a three-dimensional problem. Both have their advantages and trade-offs. The 3-D model is more accurate in its description of the process because it takes into account the spread (deformation in the direction of the thickness) that the work piece undergoes. This important factor is ignored by the 2-D plane strain models where $\varepsilon_z = 0$. The 3-D method is however more computationally expensive as the number of elements is increased. One of the most efficient approaches to the simulation of the flat imprint process is the plane-strain method. 2D model and simulation reduce the computing time and computer recourses.

Many researchers are used simplified a two-dimensional plain strain model [28, 31, 33, 40]. A small fillet was made on the mold, so as to avoid some impractical damage of the elements, such as molds penetrate through the element [31, 33, 40] (Fig. 2.40).

If the pattern of the mold is regular and symmetric just part of the mold with symmetric boundary conditions could be used for simulation [28]. Some researches are used in hot imprint process concave and convex mold type [30].

The success and efficiency of a nonlinear material solution depends on the choice of material model. Even the most complex material models are still significant idealizations of the real situation. The material nonlinearities in the hot imprint process are due to the high degree of plasticity. The plastic behavior is defined the following:

Yield Criteria: when plastic behavior is expected need tell the solver which criteria to seek to initiate yielding. The transition from elastic to the plastic state occurs when the stress reaches the yield point of the material. Yield criteria are a function of the stresses in the model.

The von Mises yield criterion given by the alternative forms of equation

$$\bar{\sigma} = f(\sigma_{ij}) = \sqrt{\frac{1}{2}\left(\sigma_x - \sigma_y\right)^2 + \left(\sigma_y - \sigma_z\right)^2 + \left(\sigma_z - \sigma_x\right)^2 + 6\left(\tau_{xy}^2 + \tau_{yz}^2 + \tau_{zx}^2\right)} = \sigma_Y,$$
(2.49)

$$\bar{\sigma} = f(\sigma_{ij}) = \sqrt{\left(\sigma_1 - \sigma_2\right)^2 + \left(\sigma_2 - \sigma_3\right)^2 + \left(\sigma_3 - \sigma_1\right)^2} = \sigma_Y,$$
(2.50)

$$\bar{\sigma} = f(\sigma_{ij}) = \sqrt{J_2} = \sqrt{s_1^2 + s_2^2 + s_3^2} = \frac{1}{2}s_{ij}s_{ij} = \frac{\sigma_Y^2}{3},$$
(2.51)

and it says "yield occurs when the equivalent stress (Mises stress) equals the yield stress in uniaxial tension $\sigma_Y$, i.e., $\bar{\sigma} = \sigma_Y$", $s_{ij}$—components of the deviatoric stress tensor [46–48].

**Fig. 2.40** Micro thermal imprint process and its analytical model: **a** schematic diagram of imprint process, and **b** simplified plain strain analytical model

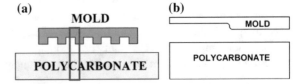

The yield surface, F, for the Drucker-Prager material law is given by

$$F = \alpha I_1 + \sqrt{J_2} = K, \tag{2.52}$$

where $I_1$ is the first stress invariant and $J_2$ is the second deviatoric stress invariant. The first stress invariant is defined using the normal stress components:

$$I_1 = \sigma_x + \sigma_y + \sigma_z. \tag{2.53}$$

The second stress invariant is defined by

$$I_2 = \sigma_x \sigma_y + \sigma_y \sigma_z + \sigma_z \sigma_x - \tau_{xy}^2. \tag{2.54}$$

The second deviatoric stress invariant can be expressed using the first and the second stress invariants:

$$J_2 = \frac{1}{3} I_1^2 - I_2. \tag{2.55}$$

If two-dimensional plane-strain conditions prevail, the Drucker-Prager criterion becomes identical to the Mohr-Coulomb criterion if the material parameters $\alpha$ and K are given by

$$\alpha = \frac{\tan \phi}{\sqrt{9 + 12 \tan^2 \phi}}, \tag{2.56}$$

$$K = \frac{3c}{\sqrt{9 + 12 \tan^2 \phi}}, \tag{2.57}$$

where c is cohesion value, $\phi$—angle of internal friction [49].

Hardening Rule: determines the material model responds to repeated stress reversals. The yield point in tension can be expected to be equal to the opposite of the yield stress in compression.

An Isotropic hardening the compressive yield always equals the tensile yield: the absolute value defined yield stress.

In order to derive the formula for hardening, total effective strain is given as

$$\varepsilon_{eff} = \varepsilon_{ep} + \frac{\sigma}{E}, \tag{2.58}$$

where $\varepsilon_{eff}$ is the effective strain and $\varepsilon_{ep}$ is the effective plastic strain. Then hardening criterion is given

$$\sigma_{hard} = \sigma_{\exp}\left(\varepsilon_{eff}\right) - \sigma_Y = \sigma_{\exp}\left(\varepsilon_{ep} + \frac{\sigma}{E}\right). \tag{2.59}$$

A kinematic hardening model will take into account the reduction in the compressive yield point after a stress reversal. Kinematic hardening is said to take place (Fig. 2.41). It is frequently observed in experiments that, after being loaded (and hardened) in one direction, many materials show a decreased resistance to plastic yielding in the opposite direction [50]. This phenomenon is known as the Bauschinger effect.

The evolution of a kinematically hardening von Mises-type yield surface (in the deviatoric plane) used to model the phenomenon is shown. The yield function for the kinematically hardening model is given by

$$\Phi(\sigma, \beta) = \sqrt{3J_2(\eta(\sigma, \beta))} - \sigma_y, \tag{2.60}$$

where

$$\eta(\sigma, \beta) = s(\sigma) - \beta \tag{2.61}$$

is the relative stress tensor, defined as the difference between the stress deviator and the symmetric deviatoric (stress-like) tensor, $\beta$, known as the back-stress tensor. The back-stress tensor is the thermodynamical force associated with kinematic hardening and represents the translation of the yield surface in the space of stresses. The constant $\sigma_y$ defines the radius of the yield surface [49].

Flow rule: establishes the incremental stress-strain relations for plastic material. The flow rule describes differential changes in the plastic strain components as a function of the current stress state.

By the Prandtl-Reuss flow rule (used for elastic plastic solid formulation) is given as

$$\dot{\varepsilon}_{ij}^p = \dot{\lambda}\frac{\partial f(\sigma_{ij})}{\partial \sigma_{ij}} = \dot{\lambda}s_{ij}, \tag{2.62}$$

$$\dot{\lambda} = \frac{1}{\tau_Y}\sqrt{\frac{1}{2}\dot{\varepsilon}_{ij}\dot{\varepsilon}_{ij}}, \tag{2.63}$$

where $s_{ij}$—components of the deviatoric stress tensor, $\dot{\varepsilon}_{ij}$—components of the strain rate tensor, $\dot{\varepsilon}_{ij}^p$—components of the plastic strain rate tensor, $\dot{\lambda}$—flow rule non-negative factor of proportionality, $f(\sigma_{ij})$—Yield function, $\tau_Y$—yield strength.

**Fig. 2.41** Elasto-plastic hardening models: *left* isotropic, *right* kinematic [51]

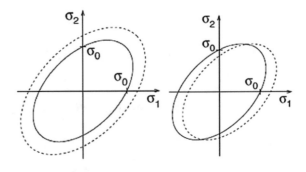

Geometric nonlinearities in the hot imprint process are due to the existence of large strains and deformations. Contact and friction problems lead to nonlinear boundary conditions. This type of nonlinearity manifests itself in several real life situations; for example, metal forming, gears, interference of mechanical components, pneumatic tire contact, and crash [52].

For the following reasons, as the specifics of the model, it is necessary to solve nonlinear deformable body of large displacements, heat transfer and contact problems.

Strain-Displacement equations describe the basic relations between displacement and strain, which are shown as Eqs. (2.64)–(2.69).

$$\varepsilon_x = \frac{\partial u}{\partial x} \tag{2.64}$$

$$\varepsilon_y = \frac{\partial v}{\partial y} \tag{2.65}$$

$$\varepsilon_z = \frac{\partial w}{\partial z} \tag{2.66}$$

$$\gamma_{xy} = \frac{1}{2}\left(\frac{\partial u}{\partial y} + \frac{\partial v}{\partial x}\right) \tag{2.67}$$

$$\gamma_{yz} = \frac{1}{2}\left(\frac{\partial w}{\partial y} + \frac{\partial v}{\partial z}\right) \tag{2.68}$$

$$\gamma_{zx} = \frac{1}{2}\left(\frac{\partial u}{\partial z} + \frac{\partial w}{\partial x}\right) \tag{2.69}$$

Here $\varepsilon_x, \varepsilon_y, \varepsilon_z, \gamma_{yz}, \gamma_{xz}, \gamma_{xy}$ are linear deformation by x, y, z direction and shear deformations on yz, xz, xy plane, u, v, w—displacements by x, y, z direction.

Strain compatibility equation is based on the consideration of the continuum assumption. When we analyze the strain-displacement behavior of a small 3-D element, we must ensure the continuity of material before and after deformation. Equations (2.70)–(2.75) are first derived by Saint-Venant from the strain-displacement equation.

$$\frac{\partial^2 \varepsilon_x}{\partial y^2} + \frac{\partial^2 \varepsilon_y}{\partial x^2} = \frac{\partial^2 \gamma_{xy}}{\partial x \partial y} \tag{2.70}$$

$$\frac{\partial^2 \varepsilon_y}{\partial z^2} + \frac{\partial^2 \varepsilon_z}{\partial y^2} = \frac{\partial^2 \gamma_{yz}}{\partial y \partial z} \tag{2.71}$$

$$\frac{\partial^2 \varepsilon_x}{\partial z^2} + \frac{\partial^2 \varepsilon_z}{\partial x^2} = \frac{\partial^2 \gamma_{xz}}{\partial x \partial z} \tag{2.72}$$

$$2\frac{\partial^2 \varepsilon_x}{\partial y \partial z} = \frac{\partial}{\partial x}\left(-\frac{\partial \gamma_{yz}}{\partial x} + \frac{\partial \gamma_{xz}}{\partial y} + \frac{\partial \gamma_{xy}}{\partial z}\right) \tag{2.73}$$

$$2\frac{\partial^2 \varepsilon_y}{\partial x \partial z} = \frac{\partial}{\partial y}\left(\frac{\partial \gamma_{yz}}{\partial x} - \frac{\partial \gamma_{xz}}{\partial y} + \frac{\partial \gamma_{xy}}{\partial z}\right) \tag{2.74}$$

$$2\frac{\partial^2 \varepsilon_z}{\partial x \partial y} = \frac{\partial}{\partial z}\left(\frac{\partial \gamma_{yz}}{\partial x} + \frac{\partial \gamma_{xz}}{\partial y} - \frac{\partial \gamma_{xy}}{\partial z}\right) \tag{2.75}$$

In summary, to obtain the stress distribution by finite element method (FEM), firstly the force load is transferred to strain load, then displacement load by constitutive equation and strain-displacement equation, respectively. Then, the continuity equations and strain compatibility equations are solved by finite element method (convert differential equation to linear equation group) and calculate the displacement for each element. Finally, the strain and stress for each element are derived from displacement data by strain-displacement equation and constitutive equation in sequence [29].

Different processes of heat transfer can take place in one or more of the following methods: thermal conduction, convection and radiation.

In hot imprint process, on top of the material and mold, there is additional heat loss to the environment due to radiation and convection, heat loss by conduction to the mold, and temperature change.

Heat transfer conductivity is described according to the formula:

$$\rho(T)c_p(T)\frac{\partial T}{\partial t} + \nabla(-k\nabla T) = q, \tag{2.76}$$

where k—thermal conductivity, $\rho$—density, cp—heat capacity, T—temperature, q —rate of the heat generation.

Contact is a non-linear boundary value problem. During contact, mechanical loads and sometimes heat are transmitted across the area of contact. If friction is present, shear forces are also transmitted.

Boundary contact conditions cause huge difficulties and make the convergence of the model extremely difficult. It is necessary to match the initial geometric shape as well as possible, but also to introduce test to determine when a node comes into contact with a rigid or elastic tool. This can be done geometrically, and the node then restored to the surface if it has apparently crossed the boundary. It is then necessary to determine whether the normal force has become tensile, before re-sitting the node [52, 53].

Comsol Multiphysics realized that the contact interaction forms: node to node, node to surface, surface to surface. FEM analysis is preferred that the surface of the deformable body share a common node with the rigid body.

When modeling contact, structural parts that come into contact have to be defined and consisted of two sets of boundaries, a slave and a master domain. The slave boundaries can't penetrate the master boundaries [54].

Describing of the contact, the polymer is as deformable body, that mechanical characteristics of the parameters is much little than the rigid body. Simulation assumed that the solid body is non-penetrated.

COMSOL Multiphysics solves contact problems using augmented Lagrangian method. This method is a combination of penalty and Lagrange multiplier methods. It means a penalty method with penetration control. The system is solved by iteration from the determined displacement. These displacements caused by incremental loading, are stored and used to deform the structure to its current geometry. If the gap distance between the slave and master boundaries at a given equilibrium iteration is becoming negative, (the master boundary is penetrating the slave boundary), the user defined normal penalty factor $p_n$ is augmented with Lagrange multipliers for contact pressure $T_n$

$$T_{np} = \begin{cases} T_n - p_n g & if\ g \leq 0 \\ T_n e^{-\frac{p_n g}{T_n}} & otherwise \end{cases}, \tag{2.77}$$

g is the gap distance variable between slave and master boundary [54, 55].

Polymers can be classified in standard polymers, technical polymers and high performance polymers, according to their temperature stability. The group of thermoplastic polymers is split into the groups of amorphous and semicrystalline polymers. Further the thermoplastic polymers can be modified with certain fillers in the micro and nano range to improve the mechanical stability and to reduce anisotropic behaviour in the molded part. Here the size and the shape of the filler in comparison to the size of the microcavities will limit the use of filled polymers.

Amorphous polymers like PMMA and PC are well suited in thermal imprint process because of their molecular structure [34]. It is known that the material property of amorphous polymer is strongly dependent on the conditions, such as temperature and loading. The material deformation behavior of amorphous polymer resists is a function of temperature can be classified into three states: glassy state, rubbery state and flow state [41]. When the temperature is lower than the glass transition temperature $T_g$, it is glassy state, in which amorphous polymer acts like a hard and brittle solid glass. The deformation is reversible. As the temperature is increased, it goes to a rubbery state, in which the polymer acts like an incompressible or approximately incompressible rubber. From glassy state to rubbery state there is no strict temperature, "jumping point", however, there is a transition region. The temperature range of this transition region is from 5 to 20 °C, depending on the characteristic of the polymer [56]. From the transition region to the rubbery state, the polymer shows dual response of reversible and irreversible deformation for mechanical stress. This property is regarded as the so-called viscoelasticity. When the temperature is further increased, it goes to the flow state, in which the polymer melts and can be regarded as a viscous non-newtonian liquid. The material deformation is irreversible [41, 47].

## 2.3.2   Finite Element Model of Hot Imprint Process

The mathematical model of the process of mechanical hot imprint into polycarbonate near it's glass transition temperature is presented in this chapter. Experimental results were compared with hot imprint simulations in order to investigate the effectiveness of the new model. These simulations help better to understand the mechanical hot imprint process. The modelling and simulation methodology by FEM including geometrical modeling, boundary conditions, meshing, material properties, process conditions and governing equations are presented. FE model was created using Comsol Multiphysics software.

Hot imprint model creation diagram is shown in Fig. 2.42.

The equations of motion, thermal balance, material properties and material deformations were used to calculate stress, strain, temperature fields, distribution of mold pressure and filling ratio in each step of the hot imprint process.

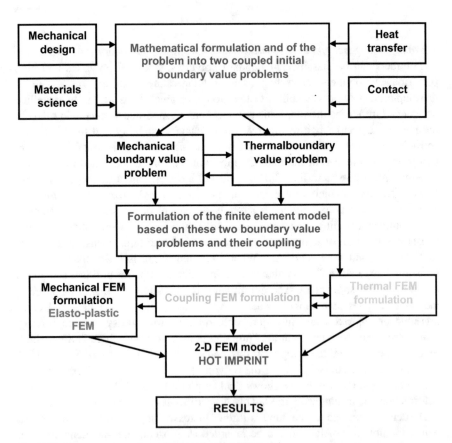

**Fig. 2.42** Diagram of creation of the hot imprint FE model

As discussed in the Sect. 2.3.1 most of the authors of scientific papers divide mechanical hot imprint process into four major steps:

1. Preheating;
2. Imprint;
3. Cooling;
4. Demolding.

In this modeling (Fig. 2.42) the elasto-plastic material model was chosen. This model is simplified by assuming, that polycarbonate can be cooled in a very short period of time, thus a separate cooling step is not analyzed.

In Sect. 2.3.1 analyzed dependence between temperature and pressure from time is modified for this model as shown in Fig. 2.43. This model consists from three steps: heating, imprinting and demolding (Fig. 2.44).

Although many periodical nanometer-scale patterns are defined on the mold, it is impractical to consider all of them using the FEA. If the cross-sectional shape of the mold is constant in one direction, as in the line pattern shown in Fig. 2.45,

**Fig. 2.43** Diagram of mechanical hot imprint process

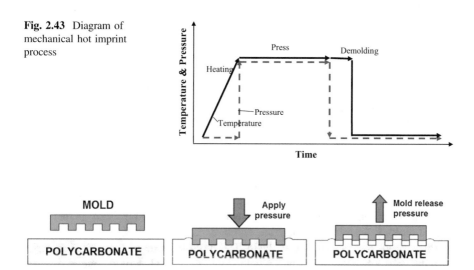

**Fig. 2.44** Steps of mechanical hot imprint process of periodical microstructure: *left* heating step; *middle* hot imprint step; *right* demolding step

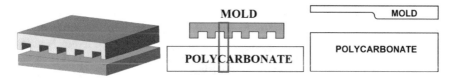

**Fig. 2.45** Geometry of the hot imprint model: *left* 3D model; *middle* 2D model; *right* parts of 2D mode

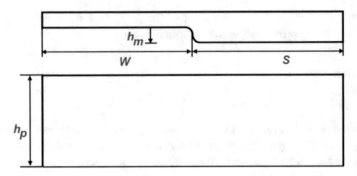

**Fig. 2.46** Geometrical parameters of the microstructure model

**Table 2.3** Model's geometrical parameters

| Parameters | Value |
|---|---|
| $h_m$—depth of the mold | 100 nm |
| $h_p$—thickness of the polycarbonate | 3 mm |
| 2W—length of the cavity | 2 μm |
| 2S—length of the ridge | 2 μm |

two-dimensional stress analysis is possible. Moreover, if the offset and recessed pattern of the mold is regular and symmetric, we can assume a two-dimensional plane strain model of a unit cell of the patterned mold using FE analysis model by taking into account symmetric boundary conditions (Fig. 2.46 and Table 2.3). To avoid stress concentration in a finite element node, a small radius was made in the corner of the mold.

The complete simulation requires two analytical models: thermal analysis, using the heat transfer module and mechanical analysis, using module of structural mechanics together with data and solutions. The boundary conditions and initial values in Comsol Multiphysics software are described in three levels: sub-domains, edges and points. Figure 2.47 shows two-dimensional (2-D) FEM model of a Nickel mold, Polycarbonate substrate and boundary conditions.

Plain strain was chosen for analysis. The mold is made of more rigid material than the polycarbonate, thus it was assumed that the mold has rigid contact surfaces in FE simulation. Symmetric boundaries were used on two sides of the model. A symmetric boundary indicates that displacement and temperature gradients across the boundary are equal to zero. Fixed normal displacement was applied to the bottom surface of the substrate. Thermal boundary and fixed normal displacement were used in order to approximate mechanical and thermal boundaries of the portion of polycarbonate's surface to the side of cavity (the thickness of the polycarbonate is much larger than thickness of cavity). Arc of small radius between mold's A and B areas is formed, in order to improve the convergence of the simulation. The initial temperatures of mold and polycarbonate are 293 K. Mold's temperature $T = f(T, t)$ is defined as function of mold's heating temperature T (Kelvin) and time t (seconds) (Fig. 2.48, Tables 2.4 and 2.5).

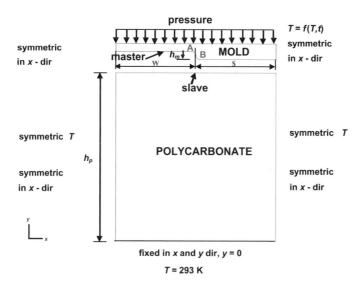

**Fig. 2.47** Boundary conditions of the hot imprint process

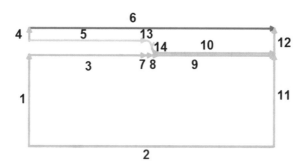

**Fig. 2.48** Models edges numeration

**Table 2.4** Mechanical boundary conditions

| Constrain | | |
|---|---|---|
| Symmetry plane | Free | Prescribed displacement |
| 1, 4, 11, 12 | 3, 5, 7, 8, 9, 10, 13, 14 | $x = 0, y = 0$   2, |
| | | $x = 0, y = y(t)$   6 |

**Table 2.5** Thermal boundary conditions

| Thermal boundary conditions | | | |
|---|---|---|---|
| Insulation/symmetry $-n(-k\nabla T) = 0$ | Temperature $T = T_0$ | Temperature $T = f(T_s(t))$ | Convective flux $-n(-k\nabla T) = 0$ |
| 1, 11 | 2 | 4, 5, 6, 10, 12, 13, 14 | 3, 7, 8, 9 |

**Table 2.6** Contact pairs

| Master | Slave |
|---|---|
| 5, 10, 13, 14 | 3, 7, 8, 9 |

The mathematical model for heat transfer by conduction is expressed by the heat equation:

$$\rho c_p \frac{\partial T}{\partial t} - \nabla(k\nabla T) = 0, \tag{2.78}$$

where k—thermal conductivity, ρ—density, cp—heat capacity, T—temperature.

In order to obtain the solution for stress, strain, rate of strain, velocity and temperature fields within the imprinted polycarbonate, appropriate boundary and initial conditions must be defined.

The contact's type in this model is rigid-deformable. The mold is rigid body (master) and polycarbonate—deformable body (slave). Since the model is a 2-D, contact boundary conditions are described in pairs of contact edges Table 2.6.

Materials, used for the mold and substrate, together with their properties are listed in Table 2.7. Nickel was used as a mold material, and it was assumed to be isotropic and elastic. The material, used as substrate, was polycarbonate. It is amorphous polymer with a glass transition temperature of about 423 K.

Simulation of hot imprint process using coupled time-depend thermo-mechanical analysis is presented in diagram (Fig. 2.49). It includes heat transfer, structural mechanics and contact analysis. Also it takes into account all necessary material parameters such as thermal conductivity (k in W/mK), density (ρ in kg/m$^3$), heat capacity (cp in J/kg K), Young's modulus (E in N/m$^2$), Poisson's ratio (v) and thermal expansion coefficient (α in K$^{-1}$), which is defined as function of temperature [42, 57].

Mechanical properties and material behaviour of the formable material during hot imprint process are extremely important when identifying optimal process conditions for the manufacture of defect-free nano-structures. An accurate determination of critical material parameters below or above $T_g$ is considered as a key requirement for the numerical simulation.

**Table 2.7** Material properties of the mold and polycarbonate

| | Mold | Substrate |
|---|---|---|
| Material | Nickel | Polycarbonate |
| Density, kg/m$^3$ | 8.908 × 103 | ρ(T) |
| Thermal conductivity, W/m K | 90.9 | k(T) |
| Thermal expansion, 1/K | 13.4 × 10$^{-6}$ | 6.5 × 10$^{-5}$ |
| Heat capacity, J/kg K | 445 | cp(T) |
| Elastic modulus, GPa | 200 | E(T) |
| Poisson's ratio | 0.31 | v(T) |

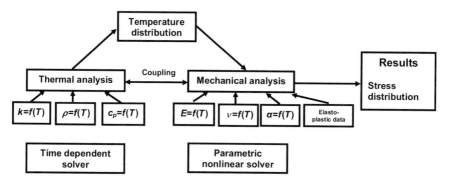

**Fig. 2.49** Diagram of thermo-mechanical analysis

The following set of graphs, which is presented by Comsol Multiphysics 3.5a material library, describes the thermal conductivity, density, Poisson ratio, heat capacity and Young modulus of the polycarbonate, which is used in hot imprint process.

In Fig. 2.50 it can be seen how thermal conductivity depends on the temperature. This is an important fact for the hot imprint process. In this process PC temperature increases rapidly, this increment significantly influences on the heat transfer between the mold and the PC.

Elastic-plastic materials are employed in order to describe deformation under large strain. In this model, polycarbonate is described as elasto-plastic material.

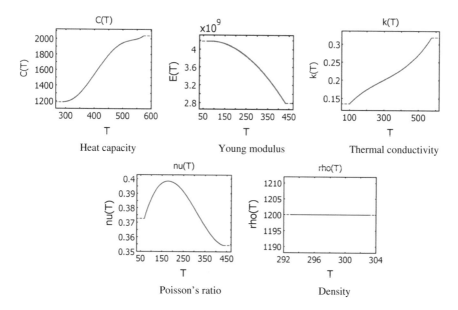

**Fig. 2.50** Thermal dependencies of polycarbonate material parameters

The micro hot imprint process is being performed near the glass transition temperature of polycarbonate, at which it behaves like elasto-plastic material. Under the constant load material would undergo two stages of deformation: an instantaneous elastic deformation $\varepsilon_e$ at the beginning of the process, followed by plastic deformation $\varepsilon_p$. The total deformation is described as:

$$\varepsilon = \varepsilon_e + \varepsilon_p. \qquad (2.79)$$

The behaviour of elasto-plastic material can be described as a model with a set of spring and friction components. Rheological models of elasto-plastic body with hardening are given in Figs. 2.51 and 2.52, where $\sigma_k$—yield strength, $E_2$—modulus of plasticity, $E_1$—modulus of elasticity (Fig. 2.51).
In Fig. 2.52 means [49]:

$$tg\alpha = E_1, \quad tg\beta = \frac{E_1 \cdot E_2}{E_1 + E_2}. \qquad (2.80)$$

Elastic-plastic models in FE systems require data in the form of elastic constants in order to describe elastic behavior and parameters, which describe yield, hardening and flow behavior in order to describe plastic behavior.

**Fig. 2.51** Schematic diagram of elasto-plastic model

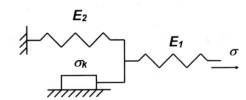

**Fig. 2.52** Deformation characteristic

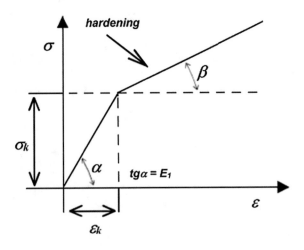

In this model the yield function von Mises was used

$$\sigma_Y = \sqrt{\frac{1}{2}(\sigma_1 - \sigma_2)^2 + (\sigma_2 - \sigma_3)^2 + (\sigma_1 - \sigma_3)^2}, \tag{2.81}$$

where $\sigma_1, \sigma_2, \sigma_3$ are principal stresses and $\sigma_Y$ is the equivalent stress.

The yield stress level $\sigma_{ys}$ and isotropic hardening $E_{Tiso}$ are as a function depending on temperature:

$$\sigma_{ys} = f(T), \tag{2.82}$$

$$E_{Tiso} = g(T). \tag{2.83}$$

The hardening function is a function of the effective plastic strain and describes the behavior, which starts from the yield stress of the material.

The geometrical singularity can cause a high rate polymer deformation. The accuracy and convergence of the solution depend on the choice of the meshing. As can be seen in Fig. 2.53, the contact areas have more refined elements than those areas, which do not have contact during the deformation. Also the mesh at the symmetrical region is carefully structured and has more refined element than the rest in domain. The model consists of 5583 finite elements.

The mesh of the model, using triangle quadratic Lagrange finite elements, is presented in Fig. 2.53. It is fine in the contact between the mold and polycarbonate. Fine mesh guarantees the convergence of the solution. The triangular element is defined by six nodes, each having three degrees of freedom: displacement in the nodal horizontal x and vertical y directions, and temperature. Finite element of Lagrange-Quadratic type was chosen. This type is used for 2-D modeling of solid structures.

**Fig. 2.53** Finite element mesh

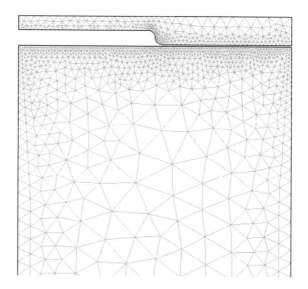

There are six nodes in this element: three corner nodes and three mid-side nodes. The displacements (u, v) are assumed to be quadratic functions of (x, y),

$$u = b_1 + b_2 x + b_3 y + b_4 x^2 + b_5 xy + b_6 y^2$$
$$v = b_7 + b_8 x + b_9 y + b_{10} x^2 + b_{11} xy + b_{12} y^2,$$

(2.84)

where $b_i$ $(i = 1, 2, \ldots, 12)$ are constants.

Several assumptions are made in order to simplify the model. The mechanical behavior of both nickel and polycarbonate at the simulated scale is governed by the equation of continuum mechanics, in which we consider all object to be continuous. The surfaces of contacting bodies are perfectly smooth. Adhesion is neglected. The cooling step is integrated into demolding step. The contact friction coefficient is assumed to be zero at the interface between the mold and polymer. No air is trapped inside the microcavities and buckling phenomenon, due to surface tension, is ignored. Heat losses to the environment are neglected. The mold remains hot all the time.

### 2.3.3   Hot Imprint Process Simulation Results

In general, the simulation model and numerical results provide with useful understanding of the fundamental formation mechanism during the hot imprint process and serve as a useful guide for specifying the optimal processing conditions for variety of hot imprint applications.

Hot imprint process was divided into three steps: heating, imprinting and demolding. In this case:

(1) Heating. The initial temperature of the mold and polycarbonate is 293 K (environmental temperature). When the stamp touches polycarbonate, the heating of the mold up to 421 K temperature begins. During the heating process, the heat is carried to the polycarbonate and it starts to deform due to the effect of the heat. The heating step lasts about $t = 2 \times 10^{-7}$ s.

(2) Imprinting. During this process, the mold goes down and presses polycarbonate, at the same time the contact force between the mold and polycarbonate increases. Polycarbonate is being further deformated and plastic deformation appears.

(3) Demolding. In this step, the hot mold $(T = 421\,\text{K})$ is demolded and finally polycarbonate is cooled. Polycarbonate assumes the form of the mold's periodic microstructure.

The model is solved using heat transfer and solid stress-strain application modes with thermal contact problem between mold and polycarbonate. This multiphysics polycarbonate hot imprint model includes the heat transport, structural mechanical stresses and deformations, resulting from the temperature distribution. It allows evaluate temperature distributions and stresses in the polycarbonate during hot imprint process. Obtained theoretical results were compared with experimental.

The imprint pressure is one of the main parameters and has a major impact on the quality of the replication. Insufficient pressure would result in incomplete filling of the pattern grooves and may subsequently lead to shape defect. Whereas too high imprint pressure causes high residual stresses in the polymer during the subsequent process step.

In Fig. 2.54 the dependence of imprint force from mold displacement is presented. The negative sign "–"shows that mold moves down and for analysis the absolute value is used. As shown in the graph, during the heating step the force increases slowly, this means that polycarbonate is still cold and the resistance force is high. Point A indicates the end of the heating step and from here the imprint force increases linearly up to point B—the end of imprint step. At this point the displacement of the mold is equal 0.9 μm. From this point the demolding step starts and as can be seen from the graph the imprint force decreases linearly till the point C—end of demolding step. The changes of polycarbonate behavior in each step are presented below.

1. Heating step

Created model shows how polycarbonate substrate would behave under thermal load and evaluates temperature, displacement, and a stress fields during hot imprint process. As shown in Fig. 2.55, maximum Von Mises stress (107 MPa) is located in the place where the mold's corner contacts with polycarbonate during the heating process. Blank mold cavity is partially filled with heated polycarbonate. Von Mises stress in polycarbonate reaches 107.1 MPa. Temperatural distribution of the specimen is presented in Fig. 2.55 using contour lines. Variation of temperature is in the range from 295 to 413 K.

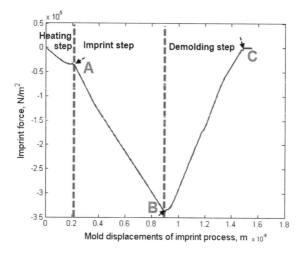

**Fig. 2.54** Imprint force dependence from mold displacement

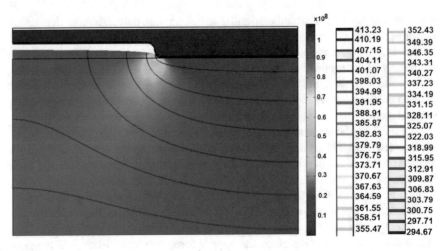

**Fig. 2.55** Von Mises stress distribution and temperature fields in the polycarbonate during heating process after $1.9 \times 10^{-7}$ s

As shown in Fig. 2.56, maximum displacement in x direction (absolute value $1.69 \times 10^{-8}$ m) is also located in the contact place between the mold's corner and polycarbonate. The sign "–" means that the direction of the displacement is to left side. The arrows demonstrate the direction of total displacement. Then the mold moves down and the polycarbonate from the right side moves to the left, as shown by arrows.

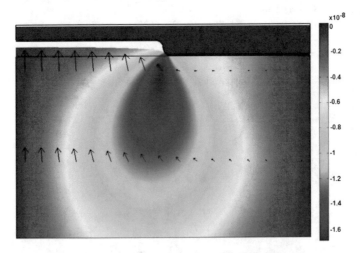

**Fig. 2.56** Displacements in x direction distribution and total displacements by *arrows* after $1.9 \times 10^{-7}$ s

In this model displacement in horizontal and vertical directions of the polycarbonate in contact edge (Fig. 2.57), in each hot imprint process step was analyzed.
Heating process lasts only $3 \times 10^{-7}$ s (Fig. 2.58). At the beginning of this process
$(0 < t < 1.5 \times 10^{-7})$ the integral displacement in x direction is equal to zero,
because at this time the temperature of the polycarbonate is still near the initial
temperature. The shifts of the polycarbonate were observed from $1.5 \times 10^{-7}$ s to
$2.7 \times 10^{-7}$ s. Polycarbonate is elastic, it means, that polymer after the contact with
mold areas shifts to the empty cavity of the mold. The cavity of the mold is partially
filled with heated polycarbonate. At the time of $2.7 \times 10^{-7}$ s deformations of the
polycarbonate stop. It means the heating process becomes steady.

2. Imprinting step

After the imprinting process (Fig. 2.59) the absolute value of the largest displacement in the x direction is observed in the empty cavity of the mold and in
contact place with polycarbonate. During the imprint process, the mold moves
down from the initial position by approximately 710 nm. Absolute value of the
polycarbonate displacement in the x direction is from 9 to 85 nm. The arrows show
that total displacement after hot imprint process is diverted to down. It means, that
polycarbonate moves down uniformly in all area.

As shown in Fig. 2.60, maximum Von Mises stress (344 MPa) is located in the
place where the mold's corner and polycarbonate contact. Blankmold cavity is filled

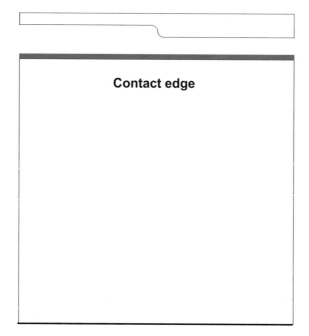

**Fig. 2.57** Contact edge in the model

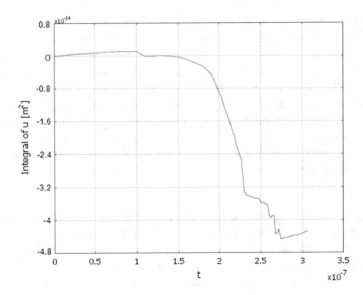

**Fig. 2.58** Integral displacement in x direction of the contact edge in heating step

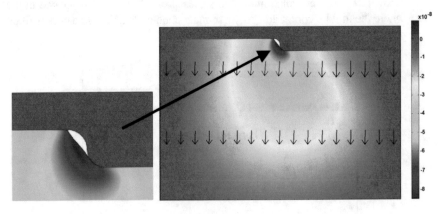

**Fig. 2.59** Deformed polycarbonate and displacement in x direction distribution and temperature fields after imprint step

with heated polycarbonate. Only small cavity remains empty. The temperature in the upper layer of the polycarbonate reaches the maximum value of 419 K. The lines of the temperature fields in the sample attain the form of the periodic microstructure. It means that there is an empty area in the mold of the polycarbonate. It appeared due to features of polycarbonate (Fig. 2.61).

During the imprint step ($0 < t < 1.4 \times 10^{-7}$ s) the integral displacement module increases linearly in x direction (Fig. 2.6). After the imprint step an empty area remains between the mold and polycarbonate (Fig. 2.59).

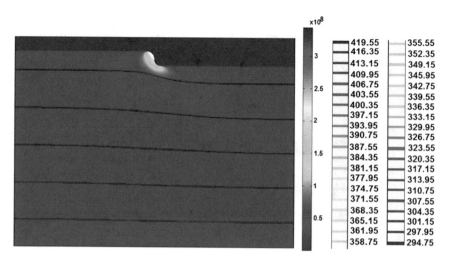

**Fig. 2.60** Von Misses stress distribution of deformed polycarbonate and temperature fields at $9 \times 10^{-7}$ s after imprint step

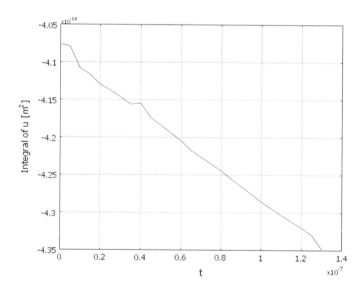

**Fig. 2.61** Integral displacement in x direction of contact edge in imprint step

3. Demolding step

When the mold was demolded and the temperature became equal to the ambient temperature 293 K, the plastic deformations took place in the polycarbonate. Their distribution is shown in Fig. 2.62. In this picture we can see how many times and in what areas they have exceeded the permissible limit of the yield.

**Fig. 2.62**  Areas of the permissible of yield

After demolding and cooling processes plastic deformation in the polycarbonate was observed. Model allows determine areas of the elastic and plastic deformations in the polycarbonate. The distribution of the ratio of Von Mises stress with yield strength of the polycarbonate is presented in Fig. 2.62. It shows that in some points Von Mises stress exceeds yield strength more than 5 times. Maximal ratio was observed in the corners of the formed structure.

In the demolding step the integral displacement module in x direction increases step by step (Fig. 2.63). It is so, because mold goes up discreetly.

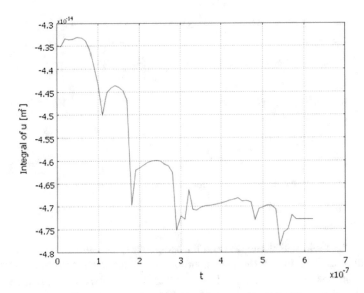

**Fig. 2.63**  Integral displacement in x direction of contact edge in demolding step

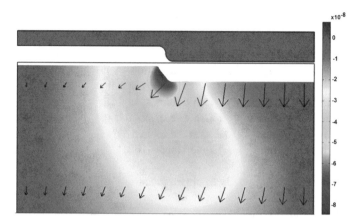

**Fig. 2.64** *Horizontal* direction displacement fields at $7.1e^{-7}$ s, *arrows* represent total displacements in PC after demolding

As can be seen in Fig. 2.64, after the demolding, displacement fields in the x direction remain the biggest at the corner of the mold. The total displacement arrows are directed down, this means that during the demolding step the mold moves up and the polymer distributes into the empty cavities.

One of the most important quality parameters in hot imprint process is the percentage of filling ratio. It is defined as ratio of filled area and all area in microstructure.

As shown in Fig. 2.65, polycarbonate empty cavity decreases slowly at end of the heating step (mold moves to $2e^{-7}$ m)—70%, then empty cavity rapidly decreases(when the mold moves to $2.3e^{-7}$ m)—10%. At the end of imprint step empty cavity is reduced till 2% but in demolding step the empty cavity slightly increases and remains about 4%.

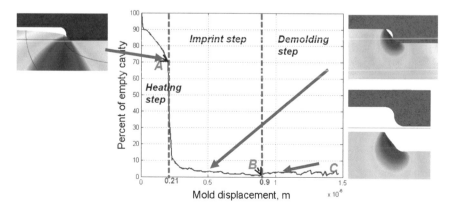

**Fig. 2.65** The dependence of non filling cavity from mold displacement

## 2.3.4  Finite Element Model Verification

The hot imprint experiment was performed on polycarbonate in order to check the validity of the results of imprint model simulation. A lamellar microstructure of 4 μm period and 100 nm depth was used in hot imprint process (Fig. 2.66). It was imprinted into polycarbonate of 3 mm thickness at 148 °C temperature, 15 s time and 5 Atm pressure. This set (temperature, time and pressure) of parameters was experimentally determined by obtaining the replica of best quality. The experiments were made in these intervals: temperature—100 to 150 °C, time—5 to 15 s, pressure—1 to 5 Atm.

It is impossible experimentally to obtain the similar dependence as shown in the model results in Fig. 2.65. The initial microstructure (nickel mold) and obtained results were compared on the basis of AFM measurements.

Polycarbonate replication results are shown in Fig. 2.67. Figures 2.66 and 2.67 represent data, which were scanned with Matlab package and interpolated nearest neighbor method. The empty area of polycarbonate is about 10%.

This shows that the difference in comparison with simulation results is approximately 2.5 times. So we can say that the model is well made and is suitable for analysis and evaluation of the material behavior.

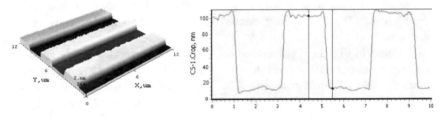

**Fig. 2.66**  Microstructure of nickel: *left* 3D surface; *right* profile

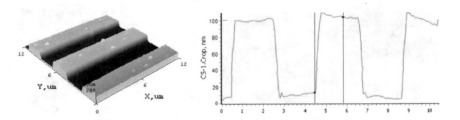

**Fig. 2.67**  Polycarbonate: *left* 3D surface; *right* profile from AFM

# References

1. Brown GC, Pryputniewicz RJ (1998) Holographic microscope for measuring displacements of vibrating microbeams using time-average electro-optic holography. Opt Eng 37:1398–1405
2. Pryputniewicz RJ, Furlong C, Brown GC, Pryputniewicz EJ (2001) Optical methodology for static and dynamic measurements of nanodisplacements. In: Proceedings of international congress on experimental and applied mechanics in emerging technologies, Portland, OR, pp 826–831
3. Pryputniewicz RJ, Stetson KA (1989) Measurement of vibration patterns using electro-optic holography. In: Proceedings of SPIE, vol 1162
4. Ostasevicius V, Palevicius A, Daugela A, Ragulskis M, Palevicius R (2004) Holographic imaging technique for characterization of MEMS switch dynamics. In: Varadan VK (ed) Proceedings of SPIE, vol 5389. Smart structures and materials 2004: smart electronics, MEMS, BioMEMS, and nanotechnology, pp 73–84
5. Ragulskis M, Palevicius A, Ragulskis L (2003) Plotting holographic interferograms for visualization of dynamic results from finite-element calculations. Int J Numer Meth Eng 56:1647–1659
6. Joroslavskij L (1987) Numerical processing of signals in optics and holography. 1987 M.: Radio i svjaz, 295 p
7. Palevičius A, Ragulskis M, Palevičius R (1998) Wave mechanical systems (theory, holographic interference). Caritas, Kaunas, 150 p
8. Palevičius A, Ragulskis M (1996) Holographic interference method for investigation of wave transport system. In: 2nd international conference on vibration measurement by laser techniques. Ancona, Italy, 1996, pp 21–27
9. Palevičius A, Ragulskis M (1998) The system of wave transportation and their holographic research. In: 3rd international conference on vibration measurement by laser techniques. Ancona, Italy, 1998, pp 125–128
10. Palevičius A, Ragulskis M, Tomasini E (1999) Vibramotor optimisation using laser holographic interferometry. In: Proceedings of 17th international modal analysis conference. Kissimme, USA SAE, 1999, pp 1012–1016
11. Gale MT (1997) Replication technology for holograms and diffractive optical elements. J Imag Sci Technol 41(3)
12. Lee B, Kwon M, Yoon J, Shin S (2000) Fabrication of polymeric large-core waveguides for optical interconnects using a rubber molding process. IEEE Photon Technol Lett 12:62–64
13. Siebel U, Hauffe R, Petermann K (2000) Crosstalk-enhanced polymer digital optical switch based on a W-shape. IEEE Photon Technol Lett 12:40–41
14. Oh M, Lee M, Lee H (1999) Polymeric waveguide polarization splitter with a buried birefringent polymer. IEEE Photon Technol Lett 11:1144–1146
15. Stutzmann N, Tervoort T, Bastiaansen C, Feldman K, Smith P (2000) Solid-state replication of relief structures in semicrystalline polymers. Adv Mater 12:557–562
16. Margelevičius J, Grigaliūnas V, Juknevičius V (1997) Forming specialities of micro-optical surfaces. In: Materials science (Medziagotyra), ISSN 1392-1320, Kaunas: Technologija 1 (4):35–37
17. Meeder M, Zehnder R, Debruyne S, Faehnle OW (2003) In-process surface measurement of replication material during UV curing. FISBA Optik AG, Rorschacher Str. 268, 9016 St. Gallen, Switzerland
18. Schulz H, Scheer H-C, Hoffmann T, Sotomayor Torres CM, Pfeiffer K, Bleidiessel G, Grutzner G, Cardinaud C, Gaboriau F, Peignon M-C, Ahopelto J, Heidari B (2000) New polymer materials for nanoimprinting. J Vac Sci Technol B 18:1861–1865
19. Guobiene A, Cyziute B, Tamulevicius S, Grigaliunas V (2002) The evaluation of diffraction efficiency of optical periodic structures. Mater Sci (Medziagotyra) 8(3):235–239
20. van Renesse RL (1998) Optical document security, 2nd edn. TNO Institute of Applied Physics Stieljesweg 1, Deft, The Netherlands. ISBN 0-89006-982-4, pp 29–55

21. Loewen EG, Popov E (1997) Diffraction gratings and applications. Marcel Dekker Inc., New York
22. Ferstl M (1998) OSA Tech Dig Ser 10:167–169
23. Martin C, Ressier L, Peyrade JP (2003) Study of PMMA recoveries on micrometric patterns replicated by nano-imprint lithography. Phys E 17:523–525
24. Baraldi      LG      (1994)      Heißpragen      in      Polymeren      fur      die Herstellungintegriert-optischerSystemkomponenten. PhD Thesis, ETH, Zurich
25. Stoyanov S et al (2011) Modelling and optimization study on the fabrication of nano-structures using imprint forming process. Eng Comput 28(1):93–111
26. Juang YJ (2001) Polymer processing and rheological analysis near the glass transition temperature. Dissertation, 230 p
27. He Y, Fu JZ, Chen ZC (2007) Research on optimization of the hot embossing process. J Micromech Microeng 17:2420–2425
28. Lan S et al (2009) Experimental and numerical study on the viscoelastic property of polycarbonate near glass transition temperature for micro thermal imprint process. J Mater Des 30:3879–3884
29. Song Z et al (2008) Simulation study on stress and deformation of polymeric patterns during the demolding process in thermal imprint lithography. J Vac Sci Technol B 26(2):598–605
30. He Y, Fu JZ, Chen ZC (2008) Optimization of control parameters in micro hot embossing. Microsyst Technol 14:325–329
31. Liu C et al (2010) Deformation behavior of solid polymer during hot embossing process. Microelectron Eng 87:200–207
32. Jeong JH et al (2002) Flow behavior at the embossing stage of nanoimprint lithography. Fibers Polym 3(3):113–119
33. Yao DG, Vinayshankar LV, Byung K (2005) Study on sequeezing flow during nonisothermal embossing of polymer microstructure. J Polym Eng Sci 45:652–660
34. Worgull M et al (2010) Hot embossing of high performance polymers. Design test integration and packaging of MEMS/MOEMS (DTIP), 5–7 May 2010, Seville, Spain, pp 272–277
35. Smidt LR, Carley JF (1975) Biaxial stretching of heat softened sheets: experiments and results. Polym Eng Sci 15(1):51–62
36. Day AJ et al (1993) Finite element modelling of polymer deformation process. In: ABAQUS user's conference, 1993, pp 151–163
37. Krishnaswamy P, Tuttle ME, Emery AF (1990) Finite element modelling of crack tip behavior in viscoelastic materials. Part 1: linear behavior. Int J Numer Meth Eng 30:371–387
38. Lin CR, Chen RH, Hung C (2002) The characterisation and finite element analysis of a polymer under hot pressing. Int J Adv Manuf Technol 20:230–235
39. Nicoli MA (2007) A thermo-mechanical finite element deformation theory or plasticity for amorphous polymers: application to micro-hot-embossing of poly(methyl methacrylate). PhD thesis, MIT
40. Kiew CM et al (2009) Finite element analysis of PMMA pattern formation during hot embossing process. In: IEEE/ASME international conference on advanced intelligent mechatronics, 14–19 July 2009. Suntec Convention and Exhibition Center, Singapore, p WB2.6
41. Kim NK, Kim KW, Sin HC (2008) Finite element analysis of low temperature nanoimprint lithography using a viscoelastic model. Microelectron Eng 85:1858–1865
42. Jin P et al (2009) Simulation and experimental study on recovery of polymer during hot embossing. Jpn Soc Appl Phys: 06FH10-1-06FH10-4
43. Hirai Y et al (2001) Study of the resist deformation in nanoimprint lithography. J Vac Sci Technol B 19(6):314–319
44. Young WB (2005) Analysis of the nanoimprint lithography with a viscous model. Microelectron Eng 77:405–411
45. Dupaix RB, Cash W (2009) Finite element modeling of polymer hot embossing using a Glass-Rubber finite strain constitutive model. Polym Eng Sci 49(3):531–543

46. Asaro RJ, Lubarda VA (2006) Mechanics of solids and materials. Cambridge University Press, New York, p 860
47. Haslach HW, Armstrong RW (2004) Deformables bodies and their material behaviour. Wiley, New York, p 560
48. Lemaitre J, Chaboche JL (1994) Mechanics of solid materials. Cambridge University Press, Cambridge, p 584
49. De Souza Neto EA, Peric D, Owen DRJ (2008) Computational methods for plasticity: theory and applications. Wiley, New York, 791 p
50. Lemaitre J, Chaboche JL (1990) Mechanics of solid materials. Cambridge University Press, Cambridge
51. Jirasek M, Bažant ZP (2002) Inelastic analysis of structures. Wiley, New York, 734 p
52. Rivera A (2007) Non-linear finite element method simulation and modeling of the cold and hot rolling process. PhD thesis
53. Crisfield MA (2001) Non-linear finite element analysis of solids and structures, vol 1. Wiley, Chichester, 345 p
54. Pennec F et al (2007) Verification of contact modeling with Comsol Multiphysics software. Paper presented at: EUROSIM: Federation of European Simulation Societies, Slovenia
55. Wriggers P (2006) Computational contact mechanics. Springer, Berlin, 518 p
56. Show MT, MacKnight WJ (2005) Introduction to polymer viscoelasticity, 1st edn. Wiley, Hoboken, 316 p
57. Reiter J, Pierer R (2005) Thermo-mechanical simulation of a laboratory test to determine mechanical properties of steel near the solidus temperature. In: Excerpt from the proceedings of the COMSOL Multiphysics user's conference. Frankfurt

# Chapter 3
# MEMS Applications for Obesity Prevention

**Abstract** Accurate measurements of the dynamics of the human body begins with the measurement data by filtering the acceleration signal evaluation taking into account the different types of human daily physical activity. Considering acceleration measuring device attached several location areas are defined on the body. The methodology of the design of micro acceleration measuring device is presented. The adequacy of accelerometer mathematical model to the physical tested experimentally using a special technique, which consists of six CCD cameras. Methodology and a special method for qualitative analysis of the human body surface tissue motion is presented. Multi level computational model assess the rheological properties of the human body surface. Interesting behavior is observed when comparing the two stages of the jump: the upper position when the velocity is zero, and the maximum speed during landing. Simulation results show that reduced surface tissue rheological model is independent of belt tension force, which is used for mounting the device. Qualitative evaluation of vertical jump, proved that the disregard of human body surface tissue rheological properties are a source of errors (up to 34%) in the analysis of human body movement.

## 3.1 Capacitive MEMS Accelerometers for Human Body Dynamics Measurements Structural Parameter Identification

New technology usually begins with the experiments. All that was ever built, must have been developed in the first place. This is immediately followed by modeling, as one wants to know how well the device works before it is built, in such a way the expensive experiments can be reduced. Modeling methods and tools enable the analysis of the existing project. The design itself is largely dependent on the experience, knowledge and creativity of the designer. Optimal synthesis methods have the potential to reduce this dependence on the human designs by automatically generating user-specified requirements [1]. Micro accelerometer synthesis algorithms have been successfully applied for the automatic marking of

© Springer International Publishing AG 2017
V. Ostasevicius et al., *Biomechanical Microsystems*, Lecture Notes
in Computational Vision and Biomechanics 24,
DOI 10.1007/978-3-319-54849-4_3

micromachined-surface accelerometers [2]. The condition for synthesis was a set of lumped parameter models that are properly connected to the device behavior using physical design variables. Concept, interdisciplinary design and optimization were introduced in [3] for the preliminary design of micro gyroscope. Optimizing the design of such a system requires a thorough understanding of the effects of the combination of working environments, physical and structural parameters of electronics, as well as processes of their manufacture. The tuning fork gyroscope was perceived as an example to demonstrate the principle of the necessary multi-disciplinary design optimization procedures in the design of MEMS. Low-g micro accelerometer, open loop single crystal silicon, has been designed and manufac-tured using the methodology of optimization [4]. Topology optimization-based approach is used in [5] for the development of micromechanical inertial power amplifiers in sensor applications. The dependence of the geometrical and mechanical parameters on the optimization is studied. The paper [6] deals with the possibility of vibration mode control for MEMS devices that have great potential in a variety of micro-sensor/actuator applications. The advantages and disadvantages arising from the use of MEMS accelerometers for hand-arm vibration and whole body measurements are estimated in [7]. The metrological characteristics of the various sensors are evaluated by determining the frequency response function, linearity, noise performance, and sensitivity to heat and electromagnetic interfer-ence. Vibration signals measured in [8] with the MEMS accelerometers are deployed tangent to the wall part, to assess the level of physical activity in the room. Accelerometers [9] are commonly used in motion analysis systems to allow researchers to conduct studies beyond traditional laboratory conditions; however, existing systems tend to be bulky and not suitable for long-term studies.

Due to the applicability of the research results it is important to clearly define how and where the device is designed to be attached on the human body. The place that is chosen must be well justified. The experiment shall be conducted to quali-tatively and quantitatively determine the influence of human body tissue surface rheology in relation to the measurement device.

### 3.1.1  Data Filtering Technique

Assume we have signal X(t) that is sampled using sampling frequency $f_s$ to obtain signal values Xi. The Discrete Fourier Transformation (DFT) can be used to acquire power spectra of the signal. However windowing must be applied in order to reduce DFT spectral leakage [10]. Hanning window was chosen as it has rapid side lobe roll-off in frequency domain and is described by following equation:

$$w(n) = 0.5 - 0.5 \cos\left(\frac{2\pi n}{N}\right)$$

(3.1)

where n = 0, 1, … , N − 1; N—number of data samples; w(n)—data sample value weighting coefficient.

Obtained signal power spectra is plotted both in normalized linear and normalized logarithmic scale plots to see and compare signal power levels at different frequencies. Noise threshold is visually picked from normalized logarithmic plot where signal levels compared to maximum signal level finishes dropping and becomes constant or is far below 0 dB (−200 to −300 dB).

Visually picked threshold is fed to an automated residual analysis process that is a quantitative tool to define possible filter pass band and cut off frequencies. Residual analysis is incremental data signal digital filtering process to obtain the relation between frequency and square differences between unfiltered and filtered data samples and can be described by following function:

$$e_r(f_f) = \frac{\sum_{i=0}^{N-1} \left| X_i - \widehat{X}_i \right|^2}{N} \tag{3.2}$$

where $X_i$ is unfiltered data, $\widehat{X}_i$—data filtered using low pass filter with pass band of ff and N—number of data samples.

For any low pass signal given function nonlinearly decreases with increasing frequency until the point where no signal information is lost due to filtering. After this point function becomes linear as further increase in frequency does not increase signal information. This turning point is minimal pass band frequency that can be plugged in low pass and differentiator filters that would filter only noise but not useful signal information.

Further step is to design and apply Finite Impulse Response (FIR) filter that takes into consideration given pass band frequency so ready to analyze signal can be obtained for further processing. FIR filters are used here for the reason that they always have a finite duration of nonzero values meaning filter stability is guaranteed [11]. The other highly desired feature is linear phase which in case of FIR filters is also guaranteed [12].

Additional step might be taken before constructing data filters. If data sampling frequency fs is high and the bandwidth B of interest is narrow compared to fs then designed FIR filters will have very high order which means that a big computational power is required for them to operate. Such problem can be dealt by using so called decimation. Decimation is the two-step process of lowpass filtering followed by an operation known as downsampling. Sequence of sampled signal values can be downsampled by a factor of M by retaining every M-th sample and discarding all the remaining samples. Relative to the original sample rate $f_{s,old}$, the sample rate of the downsampled sequence is

$$f_{s,new} = \frac{f_{s,old}}{M} \tag{3.3}$$

However, downsampling factor must be whole number that is selected in a way it would agree with Nyquist criterion:

$$f_{s,new} \geq 2B \tag{3.4}$$

where B is signal bandwidth.

Otherwise spectral aliasing will occur and the signal will be somewhat corrupted.

### 3.1.2  Methodology for Human Body Acceleration Signal Analysis

To determine the characteristics of the digital acceleration signal, and recommendations on the selection of equipment and safe filtering thresholds for use, it is necessary to analyze the body acceleration signals in order to detect the possible maximum acceleration and bandwidth of digital acceleration signal. Such an analysis can be carried out only with quantitative analysis methods such as photogrammetry, which provides accurate and reliable measurement of the image [13]. Cameras and markers were proposed for this analysis as methods with 3D constraints (space where tracking is possible) when compared with radiography or magnetic resonance imaging.

As everyday physical activities contains mostly walking and running the scope of motion for analysis to walking and running is limited. Experiment when a subject starts walking very slowly and gradually increases its speed up to the level it can still handle would fully cover the whole range of acceleration signal, which can be obtained for each day of physical activity, given the bandwidth and amplitude of the acceleration signal. However, a treadmill must be used since all activities are to be performed in a limited space where the tracking using cameras and markers is possible. Also speed range must be identified.

At the end of 2010, the highest speed of a human was recorded to be 44.72 km/h, as seen during the sprint 100 m Usain Bolt. However, such speed can be achieved only for a moment in anaerobic state (no oxygen is used to create energy in muscles). Everyday activities fall into aerobic process thus the speed range must be identified by analyzing long distance runs. According to Wikipedia, longer (over 5 km) distances running speed average is around 6 m/s (21.6 km/h), however such speed is still not sustainable for common man. Taking into consideration that average walking speed is around 5 km/h and the average speed in New York Marathon is 9.6 km/h (42.195 km in around 4 h 24 min) speed range up to 13 km/h is sufficient to cover all everyday activities.

A minimum set of points on the human body where the motion is to be tracked must be defined. This set must cover most body parts that participate in human movements since "full signal picture" is expected. Such minimum set include both wrists and tarsi where the accelerations could be the largest since legs and hands have the most level of freedom for movement. Going closer to the torso, upper arms and thighs must be tracked as an intermediate area. Finally, hips must be tracked too as they are close to the mass center of the human body and probably would produce lowest accelerations. Additional points (chest and back) must be taken to track torso movement. To sum up, the minimum set of points on the human body where the motion must be tracked in order to analyze accelerations consists of these locations (Fig. 3.1):

chest;
back;
right and left biceps;
right and left wrist;
right and left hip;
right and left thigh;
right and left tarsus.

There are two variables that are important for further decision process. Its acceleration signal bandwidth and acceleration maximum in each of defined attachment points. Acceleration maximum is required in order to select appropriate hardware and the signal bandwidth is required in order to design correct data filters to be used on the data collected by that selected hardware.

Statistical distribution of human traits is normal, only average and dispersion differs depending on the selected trait. If one test subject is taken and several measurements are done, average and dispersion can be estimated using statistical methods. By following normal distribution properties (two-sigma or three-sigma rule) it is possible to define statistically safe maximum of measured values. One might argue that one test subject is not enough to safely define parameters maximums. Thus another test subject's running and walking can be analyzed by performing the same calculations. Results then can be compared from the statistics point of view. If the distributions are considered to be the same (normal), no further test subjects are required. Thus, actual number of test subjects will be identified during the experiments.

By conducting an experiment that follows given methodology it is possible to obtain safe acceleration measurement range and acceleration signal bandwidth for everyday activities.

One more challenge is present when dealing with motion data capturing using markers and cameras. During walking and running some of the markers might get hidden from the cameras. This will break collected data integrity rendering it unusable for analysis. It is important to fill those gaps using clearly defined process to make the data fully usable. However, only small gaps can be filled. The size of the gap that can be successfully filled depends on signal sampling frequency fs and frequency f of the signal itself. The maximum number of missing digital samples that can be substituted with other values can be expressed as:

**Fig. 3.1** Set of locations to
be tracked on the human body
during the experiment

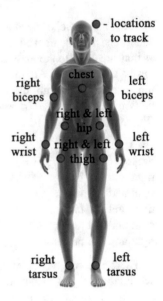

Given criteria will be used to identify if the gaps could be filled with other value. The mathematical apparatus that is selected as the tool for filling these gaps is cubic interpolation. Cubic interpolation is selected because it's not that sensitive to the high frequency noise present in unfiltered signal as the splines for example. Cubic interpolation should use at least in 0.1 s. successfully measured data points on both sides of the gap to fill the gap properly.

$$N_g = \frac{3f}{4} f_s \qquad\qquad (3.5)$$

### 3.1.3   Acceleration Measurement Device Attachment Location Considerations

Results of the analysis of the human body acceleration signals will offer some recommendations for the accelerometer to choose, as well as the device that will be developed.

However, its attachment on the body must be considered before developing the device itself as it may reveal some additional information that might be required.

There are a number of ways and places for a small device with an accelerometer to be attached on the human body in the physical exercise monitoring application.

Depending on the actual application, such devices can be worn:

- in the waist using the belt clip, or they can simply lie on the bottom of the side or back pocket (as most pedometers and activity counters worn);

- on the chest by means of special belts (common to heart rate monitors), or as part of a garment (arisen from increasing use of smart textiles);
- on the biceps using special Velcro pouches (quite common for certain types of activity monitors or GPS tracking devices);
- on the wrist, using the clock—like attachments (common for heart rate monitors and counters locomotion);
- on the thigh or the tarsus with special belts or pouches (common for running, walking, or gait monitoring devices for analysis).

Such a wide variety of mounting options evolved from different application requirements, the end user convenience and usability requirements. The waist, chest and leg are attractive places for the device to be attached in a physical activity monitoring applications, as the dynamics of these areas are strongly correlated with daily activities and normal movements such as walking and jogging. These places are also very attractive for end users, because the devices out there, as a rule, do not pose any restrictions on the movement and are comfortable to wear. Combined with new smart textile technology, these devices may be small portable systems that can be worn discreetly for an external observer, and provide a level of comfort of casual clothes. This comfort extends monitoring physical activity to new horizons, allowing physical activity to be tracked 24/7 over long periods of time.

Although there are many places where some wearable device might be attached, it is recommended to place it on the waist in the back for physical activity monitoring applications as this point is closest to the center of mass of the body and collected activity data has smallest variance.

However, attachment to the chest area is chosen because of the following:

Application of smart textiles integration into clothes as wearable systems starts to emerge.

As the final idea is to monitor daily physical activity for long periods of time and heart work is also very important information to acquire, the only place for such system not to be distributed is to put everything in one piece that is attachable to the chest.

Chest area is convenient and comfortable place for the end user to wear such system as it is unnoticeable for others.

It is concluded that the device should be attached to the chest area.

### 3.1.4   Methodology for Validation of the Operation of the Acceleration Measurement Device

When acceleration measurement range and acceleration signal bandwidth is obtained the operation of such device must be validated. It is important to validate that such device outputs correct measurements in the defined bandwidth with relatively low errors so the device can be successfully used in further research.

A dynamic excitation system can be used to validate the operation of acceleration measurement device. Such system would consist of:

Signal generator to control the excitation frequency;
Some stand that has the ability to move according to the given excitation law;
Some tracking subsystem that would capture actual movement of the stand.

Acceleration measurement device must be attached to the stand that is excited with different frequencies throughout the bandwidth of interest. Then motion data that is captured by the tracking subsystem can be compared with the data that is captured by acceleration measurement device. Such comparison of two data sources would allow making a conclusion on the operation of acceleration measurement device.

Although the validation process might seem straightforward, it is not. One of the problems that lie ahead is the attachment of the acceleration measurement device on the stand. All 3D industrial accelerometers measure acceleration vector referring to their own coordinate system (Fig. 3.2).

This means that the output signal of the acceleration measuring device is relative to the accelerometer itself but not to the global coordinate system as one might think. In a situation when the movement is free, the relation between global coordinate system and coordinate system of the accelerometer is constantly changing and that yields the measurements unusable because the reference point for each data sample is not known (is constantly changing). The reference can theoretically be tracked by integrating acceleration data to obtain the information on how accelerometer's coordinate system moves in the global coordinate system so the measurements can be transformed to have global coordinate system as the reference. However, it is practically impossible due to integration errors and measurement inaccuracies.

Another accelerometer feature is that it measures not only the acceleration that is due to its movement but also the acceleration due to gravity. This means that accelerometer will output acceleration measurement of 0 m/s$^2$ only in free fall situation or when hanging in space where no gravity is present. In all other scenarios the measurement result will be the sum of acceleration due to motion and acceleration due to gravity and this contaminates the measurements even more.

**Fig. 3.2** Coordinate system
of the 3D accelerometer

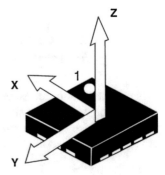

**Fig. 3.3** Accelerometer's
position in relation to global
coordinate system

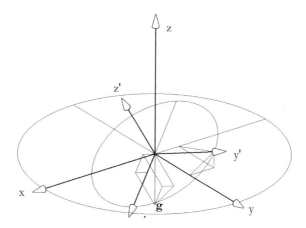

So how is it possible to overcome these problems? Well, it isn't, at least not in general case. A validation experiment must be designed in a special way that it would deal with those severe problems just described. As it was said before, the validation of acceleration measurement device is not that straight forward after all.

The movement of the stand where the acceleration measurement device is to be attached must be restricted to the uniaxial movement. This restriction can be exploited to overcome the problem of acceleration due to gravity interfering with the measurements. As it was mentioned earlier, accelerometers measure acceleration vector projections on its axes X'Y'Z' (Fig. 3.3). In a special case when it is guaranteed that coordinate system X'Y'Z' will not change its relation regarding angles to global coordinate system XYZ, mathematics to remove acceleration vector due to gravity g from measurements are described below. Such special case is obtained by restricting the stand movement to only one axis.

Let's say a number of acceleration measurement samples are taken following these two guidelines:

Sampling data for a few seconds in the steady state;
Sampling data for processing.

The average acceleration vector d of sampled time window of few seconds will not only give the magnitude but also the position of g in respect of accelerometer:

$$\mathbf{d} = \frac{\sum_{i=n}^{i=n+N} \mathbf{a}_i}{N} \tag{3.6}$$

where i is sample index, N—number of samples to be used, n—index of first sample that is included, $\mathbf{a}_i$—actual sampled vectors.

Having steady state samples is important because calculated average d is used by space transformation—rotation around some axis—as a reference vector. Then vector of length 1 on rotation axis is defined as:

$$\hat{\mathbf{o}} = \frac{\hat{\mathbf{n}} \times \mathbf{d}}{|\hat{\mathbf{n}} \times \mathbf{d}|} \tag{3.7}$$

here $\hat{\mathbf{n}} = (0; 0; 1)$.

Rotation angle cosine can be calculated from following equation:

$$\cos\varphi = \frac{\mathbf{d_z}}{|\mathbf{d}|} \tag{3.8}$$

Every subsequent sample that is measured after the window that was used to get steady average is transformed using the space rotation formula to gain a transformed data set $\langle \mathbf{a}_i^* \rangle$:

$$\mathbf{a}_i^* = \hat{\mathbf{o}}(\hat{\mathbf{o}} \cdot \mathbf{a_i}) + (\mathbf{a_i} - \hat{\mathbf{o}}(\hat{\mathbf{o}} \cdot \mathbf{a_i}))\cos\varphi + \mathbf{a_i} \times \hat{\mathbf{o}}\sin\varphi \tag{3.9}$$

here $i = \overline{0, S-1}$ where S is the number of data samples.

The equation is rearranged to remove all trigonometry functions:

$$\begin{aligned}
\mathbf{a}_i^* &= \hat{\mathbf{o}}(\hat{\mathbf{o}} \cdot \mathbf{a_i}) + (\mathbf{a_i} - \hat{\mathbf{o}}(\hat{\mathbf{o}} \cdot \mathbf{a_i}))\cos\varphi + \mathbf{a_i} \times \frac{\hat{\mathbf{n}} \times \mathbf{d}}{|\hat{\mathbf{n}} \times \mathbf{d}|}\sin\varphi \\
\mathbf{a}_i^* &= \hat{\mathbf{o}}(\hat{\mathbf{o}} \cdot \mathbf{a_i}) + (\mathbf{a_i} - \hat{\mathbf{o}}(\hat{\mathbf{o}} \cdot \mathbf{a_i}))\cos\varphi + \frac{\mathbf{a_i} \times (\hat{\mathbf{n}} \times \mathbf{d})}{|\mathbf{d}|} \\
\mathbf{a}_i^* &= \hat{\mathbf{o}}(\hat{\mathbf{o}} \cdot \mathbf{a_i}) + (\mathbf{a_i} - \hat{\mathbf{o}}(\hat{\mathbf{o}} \cdot \mathbf{a_i}))\cos\varphi + \frac{\hat{\mathbf{n}}(\mathbf{a_i} \cdot \mathbf{d}) - \mathbf{d}(\mathbf{a_i} \cdot \hat{\mathbf{n}})}{|\mathbf{d}|} \\
\mathbf{a}_i^* &= \hat{\mathbf{o}}(\hat{\mathbf{o}} \cdot \mathbf{a_i}) + (\mathbf{a_i} - \hat{\mathbf{o}}(\hat{\mathbf{o}} \cdot \mathbf{a_i}))\frac{\mathbf{d_z}}{|\mathbf{d}|} + \frac{\hat{\mathbf{n}}(\mathbf{a_i} \cdot \mathbf{d}) - \mathbf{d}(\mathbf{a_i} \cdot \hat{\mathbf{n}})}{|\mathbf{d}|}
\end{aligned} \tag{3.10}$$

Given transformation yields data transformed in the space where the plane $\mathrm{X'Y'}$ of the accelerometer matches global XY plane. It is pointed out that this only matches the planes, but not the axes themselves. Such transformation is possible due to the fact that the accelerometer's coordinate system's relation to global coordinate system regarding angles does not change.

Now it is possible to remove influence of vector g from the transformed data. To achieve this, g is simply subtracted from all samples in the set (as g and accelerometer's Z axis are parallel or in one line):

$$\mathbf{a}_i^{**} = \mathbf{a}_i^* - \mathbf{g} \tag{3.11}$$

Now data must be transformed back to its original reference system. To achieve this one needs to use this formula:

$$\mathbf{a}_i^{***} = \hat{\mathbf{o}}\big(\hat{\mathbf{o}} \cdot \mathbf{a}_i^{**}\big) - \big(\mathbf{a}_i^{**} - \hat{\mathbf{o}}\big(\hat{\mathbf{o}} \cdot \mathbf{a}_i^{**}\big)\big) \cos\varphi - \mathbf{a}_i \times \hat{\mathbf{o}} \sin\varphi$$

$$\mathbf{a}_i^* = \hat{\mathbf{o}}\big(\hat{\mathbf{o}} \cdot \mathbf{a}_i\big) + \big(\mathbf{a}_i - \hat{\mathbf{o}}\big(\hat{\mathbf{o}} \cdot \mathbf{a}_i\big)\big)\frac{\mathbf{d}_z}{|\mathbf{d}|} - \frac{\hat{\mathbf{n}}(\mathbf{a}_i \cdot \mathbf{d}) - \mathbf{d}(\mathbf{a}_i \cdot \hat{\mathbf{n}})}{|\mathbf{d}|} \qquad (3.12)$$

The MATLAB routines transform measurement data so that accelerometer's Z axis would be aligned with global axis Z.

Given methodology allows removing the interference of the acceleration due to gravity and does not require aligning stand movement axis with one of the accelerometer axes that is required for the data to be comparable. What is more, such alignment is nearly impossible, because the accelerometer is soldered on the printed circuit board which is screwed in the enclosure which is put on the stand. Geometrical relations between each of the mentioned elements are not defined. This means that relation between accelerometer's axes and stand movement axis is not known no matter what. Thus, different approach is needed to make this irrelevant for comparison.

Such approach would require mounting the acceleration measurement device without trying to align the axis of the stand movement with one of the accelerometer's axis, but rather make them clearly misaligned and use collected data to obtain that axis from the experimental results. The axis of the stand movement can be defined as:

$$\begin{cases} x = b_x + k_x t \\ y = b_y + k_y t \\ z = b_z + k_z t \\ t \in \mathbb{R} \end{cases} \qquad (3.13)$$

where kx, ky, kz, bx, by and bz are coefficients and t is any real number.

Because the acceleration measurement device is firmly attached to the stand, and the stand movement is uniaxial, it means that the relation between global coordinate system and accelerometer's coordinate system has only one degree of freedom— distance change between each along the axis of stand movement. This means that all data samples which are measured by the accelerometer will lie along the same stand movement axis (assuming no measurement errors are present). However, measurement errors are unavoidable thus the measured data samples will make sort of a cloud around the axis of the stand movement. Here, a well-known least square method can be used to calculate the axis of the stand movement [which equals to finding all the coefficients for Eq. (3.13)] by obtaining the line that best fits the experimental data.

To sum up everything that was presented it is concluded that acceleration measurements must be transformed to match references with stand movement axis so both data sets would have the same reference and can be compared. This is achieved by such process:

1. Acceleration measurements at steady state (when stand is not moving) can be used to extract accelerometers position in relation to acceleration vector due to gravity g.
2. Extracted position data is used to transform all acceleration measurement in space so accelerometer's Z axis would match vector g. Then g is subtracted from every measurement and all measurement data is transformed back to its original position. This way, acceleration measurements which are not affected by presence of acceleration due to gravity are acquired. Acquired measurements are used to extract stand movement axis data by applying least square method.
3. All acceleration data samples are rotated in space so accelerometer's Z axis would match stand movement axis.
4. Measured acceleration data is compared to stand acceleration data to evaluate accelerometer's operation reliability.

The final step would be to draw conclusions about the reliability of the tested acceleration measurement device and state whether it can be used.


### 3.1.5  Accelerometer Model and Its Validation

A system with automatic, human physical activity classification capabilities is of high demand, for applications in fields of healthcare monitoring and advanced human-machine interface development. Human physical activity data is invaluable when assessing biomechanical and physiological, variables and parameters on a long-term basis. Significant deviations in estimation may emerge when employing composite, wearable sensor systems (i.e. accelerometers) without consideration of actual nature of subjects activities [14].

To cover the acceleration signal range, observed throughout the daily physical activities of the subject, with considerations to signal bandwidth and amplitude, an experiment, observing the human walk cycle, with intensity varying from normal to highest bearable threshold, would suffice. Human body acceleration signal analysis results could provide guidelines for the choice of the accelerometer. An essential motion component in daily physical exertions is the vertical movement. Among these exertions, the most commonly encountered are: running, sitting up and down, walking. Tracking of the body areas (as per their definition in Fig. 3.1), was performed through employment of six cameras from ProReflex (MCU 500, type 170 241), using Track Manager Software (from Qualisys), the walking cycles themselves were conducted on a Vision Fitness Premier treadmill (model—T9450 HRT). The cameras employ a CCD image sensor with $680 \times 500$ pixel resolution. Use of CCD sensors, allows for considerable reduction of low noise when comparing to CMOS sensors of higher resolution. Through the use of a patented algorithm, for sub-pixel interpolation, the effective resolution of the camera becomes $20,000 \times 15,000$ subpixels, under normal operation settings, under some circumstances enabling the ProReflex MCU cameras to discern movements on the order of 50 μm [15]. Experiments were initiated with a slow 0.8 km/h walk, which incrementally increased up until peaking at an intensive run, at speed of 13 km/h, followed by a gradual return to the initial 0.8 km/h walk. The speed variation was performed in steps of 0.1 km/h, delaying each step by 5 s. Each test subject underwent three acceleration/deceleration cycles. All motion information, in all three axes, was recorded at sampling rate of 500 Hz, in order to maintain maximum allowable data resolution. Filtering of all the recorded data, was performed using a low pass, 20 Hz filter, with 80 dB attenuation at 25 Hz as bandwidth of 20 Hz—in which human movements naturally occur.

A routine was developed in MATLAB, to perform residual analysis, for each data set in the 20 Hz bandwidth. Residual analysis yields calculation of error ef, for each frequency in the range of 1–20 Hz:

$$e_f = \max(|x_i - y_i|) \tag{3.14}$$

In this case, xi is the sample that was measured, yi is the filtered sample, using f passband frequency.

Monitoring of ef values was performed with f gradually decreasing from 20 Hz, at 1 Hz increments.

A value of frequency f is considered to be the safe passband value, at the point, where ef begins to increase non-linearly, with the decrease of f. According to the analysis (Table 3.1), safe passband frequency value is equal to 16 Hz.

In order to obtain values of acceleration for each tracked marker, the differentiator filter was applied two times. The maxima of accelerations yielded during the experiments are provided in Table 3.2. Research [16] yielded tarsus (talus) area accelerations, at peak-to-peak values of 6.75 g. Results acquired during current experiments were in good agreement with [16], however, in current case, the scope of analysis, involved running at speeds of up to 13 km/h, in addition to walking. The difference in the obtained maxima values is likely due to data filtering, as high frequencies would be left out in such case.

**Table 3.1**  Residual analysis results for walking/running task data samples

| Data set | Suggested pass band frequency, Hz |
|---|---|
| Chest | 10 |
| Back | 10 |
| Right biceps | 11 |
| Left biceps | 11 |
| Left hip | 11 |
| Right hip | 10 |
| Left wrist | 10 |
| Right wrist | 10 |
| Left thigh | 16 |
| Right thigh | 16 |
| Left tarsus | 16 |
| Right tarsus | 16 |

**Table 3.2** Maximum accelerations observed in three directions

| Position | Max ($|ax|$), m/s2 | Max ($|ay|$), m/s2 | Max($|az|$), m/s2 |
|----------|--------------------|--------------------|-------------------|
| Chest    | 15.879             | 38.277             | 35.708            |
| Back     | 22.437             | 38.587             | 18.750            |
| Biceps   | 23.812             | 19.476             | 39.375            |
| Hip      | 28.145             | 37.636             | 45.645            |
| Wrist    | 31.656             | 32.347             | 71.700            |
| Thigh    | 35.736             | 36.232             | 39.749            |
| Tarsus   | 66.873             | 24.016             | 56.383            |

As stated in [17], the typical acceleration amplitude of a body can reach up to 12 g. Since availability of accelerometers, commonly used in industry is limited to ranges of ±2 g, ±4 g, ±6 g, ±8 g, ±16 g—±16 g acceleration data was chosen by default as it is closest to the ±12 g value.

According to [14], the sensing part of MEMS accelerometers tends to fit into areas of ~1 mm$^2$. Hence, the initial condition for the model is to fit the sensing element into an area of 1–1.3 mm$^2$. Another common trend in manufacturing of MEMS accelerometers is usage of silicon as the primary material [18]. Even though, the common manufacturing process develops the sensing part as a 2D structure, the requirement for the measurements to be available, at equal sensitivity, on all three axes (since it is a 3D accelerometer), must be taken into account. Hence, lumped models for the main components of the proof-mass support are not to be defined. Based on observed acceleration values, and resultant guidelines, the accelerometer should have a measuring range of ±16 g. The overall non-linearity from the sensing element and other sources such as electronics, should not exceed 1% full scale output. The bandwidth (±3 dB) of the accelerometer should be above 100 Hz, while the cross-axis sensitivity, should not exceed 1% of FSO. The values for hysteresis and bias stability of the accelerometer are each defined as 0.15% of FSO. The required response time for the accelerometer, should be lower than 1 ms and performance should be stable at temperature range between −20 and 80 °C. The capacitive accelerometer should be configured in such a manner, so that the proof-mass would avoid squeeze damping effect and would be supported on four sides by beams, shaped in a way, to enable, piston like movement, while maintaining it parallel to the electrodes at all accelerations. Additionally, any kind of longitudinal change in geometry of the beam, due to temperature fluctuations, should only limit the in-plane rotation of the proof-mass, without introducing any out of plane bending. Such configuration enables reduction of the overall chip size, thereby increasing the yield per wafer, while reducing non-linearity related to support structures, at the same time. L-shaped cantilever beams (Fig. 3.4), in this case, are the most favorable configuration. Beams of such configuration tend to provide higher sensitivity [19]. Hence, usage of the L-shaped beams as the primary structure with internal damping capability, for the sensing part model, has been added to the model requirement list.

**Fig. 3.4** Computational scheme of MEMS accelerometer

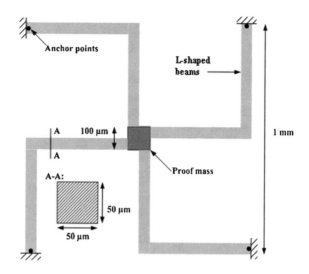

**Table 3.3** Initial accelerometer model properties

| Property | Value |
|---|---|
| Material | Si |
| Overall size (top) | 1 mm$^2$ |
| L-shaped beam cross section size | 50 × 50 μm |
| Proof mass size | 100 × 100 μm |

Taking into consideration all of the requirements, the initial accelerometer model geometry is defined and provided in Fig. 3.4 and Table 3.3.

A 3D, finite element model of the accelerometer (Fig. 3.4) was developed using Comsol and MATLAB, to enable investigation of optimization technique applicability to identifying MEMS accelerometer structure parameters [20, 21]. Since, MEMS operation requirements state, that measurement of the sensing part should be available on all three axes at equal sensitivities, the cross-section of the beam and, proof-mass material mass density were defined as the structure parameters.

Model's sensitivity equality requirement is defined in a following manner:

$$\begin{cases} u(1,A) = u(2,A) = u(3,A) \\ u(i,A) \neq 0 \\ A \in (-160,0) \cup (0;160) \end{cases} \qquad (3.15)$$

here u(i, A) is the value of displacement along axes x, y and z (i = 1, 2, 3 accordingly) at acceleration of magnitude A, being applied to the model along the aforementioned axis.

To describe the error function, the following definition is used:

$$e(i,t) = |\ddot{u}_m(i,t) - a(i,t)|$$  (3.16)

In this case i = 1, 2, 3 again, stands for the axes x, y and z accordingly; $\ddot{u}_m(i,t)$—is the acceleration exhibited by the model at certain moment t. It is assumed that governance of the dynamic excitation is defined by a (i, t) = sin(2πft), with f representing the excitation frequency.

Beam cross-section height and mass density of proof-mass material are obtained by solving the following problem:

$$\min_{t \geq 0} \left| \ddot{u}_m(i,t,h,\rho) - \frac{d^2(\sin(2\pi ft))}{dt^2} \right|$$  (3.17)

in this case $\ddot{u}_m(i,t,h,\rho)$—defines the acceleration of proof-mass at specific moment t at cross section height h, of the accelerometer beam and mass density $\rho$, of proof-mass material and f as the excitation frequency.

The optimization problem outlined above must conform to the following constraints:

$$\begin{cases} f \in (0;20) \\ u(i,A,h,\rho) = u(j,A,h,\rho), \quad i \neq j \\ u(i,A,h,\rho) \neq 0 \\ A \in (-160;0) \cup (0;160) \\ i,j = 1,2,3 \\ h \in (0;10^{-4}) \\ \rho \in (0;10^4) \end{cases}$$  (3.18)

These constraints are a result of both, earlier described requirements and real world conditions:

| | |
|---|---|
| $f \in (0;20)$ | limit of frequency (e.g. to 20 Hz); |
| $u(i,A,h,\rho) = u(j,A,h,\rho), i \neq j$ | requirement for axis sensitivity; |
| $u(i,A,h,\rho) \neq 0$ | a boundary, for avoiding zero displacement value, at which, formally, the equal sensitivity across the axes would still be met; |
| $A \in (-160;0) \cup (0;160)$ | limit of acceleration in both directions, to 160 m/s², due to range limit of ±16 g, of chosen accelerometer; |
| $h \in (0;10^{-4})$ | constraint on the height of the beam cross-section, representing real world manufacturing process boundary; |
| $\rho \in (0;10^4)$ | constraint on mass density |

According to general practice of mechanical spring usage for accelerometer design [15], the width w of the cross-section of the accelerometer is explicitly defined to be 4 μm.

Based on the origin of the problem it is determined, that a single value of A will suffice. This is because; all other values would yield identical result due to sensitivity equality across the axes requirement. Thus A is set to 10 m/s$^2$.

Therefore, the problem (3.18) can be redefined as:

$$\min_{t \geq 0} \left| \ddot{u}_m(i, t, h, \rho) - \frac{d^2(\sin(2\pi ft))}{dt^2} \right| \tag{3.19}$$

$$K(h)u + C(h)\dot{u} = -M(h, \rho)\ddot{u}$$
$$u(0) = \dot{u}(0) = 0 \tag{3.20}$$
$$F_x(t) = F_y(t) = F_z(t) = Am_a \sin(2\pi ft)$$

With constraints taking the following form:

$$\begin{cases} f \in (0; 20) \\ u(i, h, \rho) = u(j, h, \rho), i \neq j \\ u(i, h, \rho) \neq 0 \\ i, j = 1, 2, 3 \\ h \in (0; 10^{-4}) \\ \rho \in (0; 10^4) \end{cases} \tag{3.21}$$

Equation (3.20) governs motion, there C, K, M—damping, structure stiffness and mass matrices u, u̇, ü—displacement, velocity and acceleration vectors. Fx, Fy, Fz are body loads that are prescribed as external force/volume for all directions at acceleration A multiplied by a mass ma of MEMS accelerometer.

To solve the problem, a hill climbing technique (belongs to local search family) was chosen. Initial values of the optimization problem variables were h = 1 10$^{-5}$ m, ρ = 2000 kg/m$^3$. Minimum of error function was achieved when sensitivity equality was satisfied at h = 8.25 × 10$^{-6}$ m and mass density of proof-mass material of ρ = 9083.2 kg/m$^3$ while yielding minimal error function value of 5.06 × 10$^{-10}$ m. The closest density value of a corresponding real life material is ρ = 8960 kg/m$^3$, which is copper, hence, it was chosen as proof-mass material.

The final properties of the model are outlined in Table 3.4.

**Table 3.4** Final accelerometer model properties

| Property | Value |
|---|---|
| L-shaped beam material | Si |
| Proof mass material | Cu |
| Overall size (top) | 1.23 mm$^2$ |
| L-shaped beam cross section size | 4 × 8.25 μm |
| Proof mass size | 100 × 100 × 100 μm |

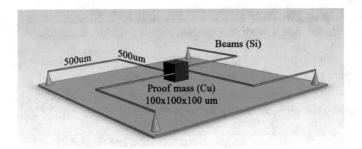

**Fig. 3.5** Final accelerometer model geometry

Model is of isotropic material. Silicon's (Si) Young's modulus is 170 GPa, Poisson's ratio—0.28 and density—2329 kg/m$^3$. Copper's (Cu) Young's modulus is 120 GPa, Poisson's ratio—0.34 and density—8960 kg/m$^3$. According to geometry specifications outlined in Table 3.4, final model (Fig. 3.5) was constructed using Comsol Multiphysics.

In order to verify that first resonant frequencies on every axis are outside the 20 Hz frequency range first three Eigen frequencies were computed. Results are given in Table 3.5. The response when acceleration of 10 m/s$^2$ was applied to the model the displacement in z direction was 50.49 nm (Fig. 3.6), displacement in y direction was 50.55 nm (Fig. 3.7) and displacement in x direction was 50.54 nm (Fig. 3.8). Difference of displacement in x direction from z direction is 0.090 and 0.006% from displacement in y direction, while displacement in y direction differs by 0.092%. According to the results it can be stated that equal sensitivity for all axes was achieved in the model.

Accelerometer model was validated using experimental data obtained by monitoring uniaxial vibration stand movement. The displacement data of vibration stand was used as an input for the model for observing base excitation law and model's response. Acceleration data obtained from the model output was compared with experimental data to evaluate the adequacy of the model.

Proof-mass displacements were analyzed for the entire acceleration range obtained from the optimization problem constraints (Eq. 3.21). Results suggest that dependence between displacement and acceleration is linear, which entails that conversion of acceleration into digital signal is possible through measurement of the capacitance between two plates, with one plate placed on the bounding box and other on the proof mass. Electrical capacity of the two plates is inversely dependent on the size of the gap between the two plates and linearly depends on the area of overlap between the plates.

**Table 3.5** Accelerometer model eigen frequencies

| First three eigen frequencies | Value, Hz |
| --- | --- |
| In x direction | 2238.81 |
| In y direction | 2239.11 |
| In z direction | 2244.52 |

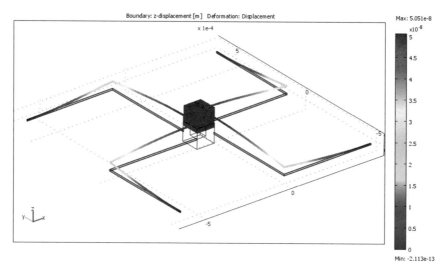

**Fig. 3.6** Displacements field in z direction when acceleration of 10 m/s² along z axis is applied

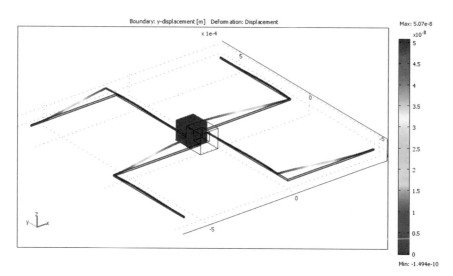

**Fig. 3.7** Displacement field in y direction when acceleration of 10 m/s² along y axis is applied

Output of the model was compared to experimental data for all excitation frequencies employed during the experiment. Table 3.6 shows the results.

According to the results, the model follows experimental data with a significant accuracy. Even though relative errors exceed 9% of the acceleration amplitude, absolute errors do not exceed 0.12 m/s². This shows that the model is valid and stable throughout the bandwidth of 20 Hz.

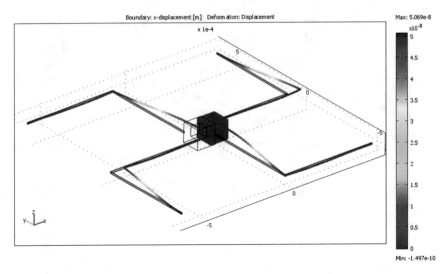

**Fig. 3.8** Displacement field in x direction when acceleration of 10 m/s$^2$ along x axis is applied

**Table 3.6** Accelerometer's model output comparison with vibration stand accelerations

| Frequency, Hz | Excitation amplitude, mm | Acceleration amplitude, m/s$^2$ | Model acceleration amplitude, m/s$^2$ | Error, % | Error, \|m/s$^2$\| |
|---|---|---|---|---|---|
| 1 | 0.8813 | 0.3480 | 0.3169 | 9.82 | 0.0311 |
| 4 | 0.9517 | 0.6772 | 0.6296 | 7.56 | 0.0476 |
| 7 | 0.9510 | 1.8397 | 1.7461 | 5.36 | 0.0936 |
| 10 | 1.0822 | 4.2725 | 4.1853 | 2.08 | 0.0872 |
| 14 | 1.0335 | 7.9967 | 7.8770 | 1.52 | 0.1197 |
| 17 | 0.9891 | 11.2849 | 11.1665 | 1.06 | 0.1184 |
| 20 | 0.9241 | 14.5920 | 14.4876 | 0.72 | 0.1044 |

## 3.2   Identification of Human Body Rheological Properties for Evaluation of the Obesity Level

### 3.2.1   Obesity as the 21st Century Catastrophy

Obesity causes a variety of physical disabilities and physiologic problems and overweight significantly increases a person's risk of developing a number of non-communicable diseases such as cardiovascular disease, cancer and diabetes. The risk of having some of these diseases (comorbidities) also increases with increasing body weight. Obesity is already responsible for 2–8% of health care costs—in accordance with [22], they can reach 10.4 billion euros—and 10 to 13% of deaths in different parts of the European region. WHO estimates that in 2020 inappropriate lifestyle lacking physical activities will be the main factor of over

70% of all diseases [23]. Relationships between physical activity and mortality, cardiovascular diseases and 2nd type diabetes spread are also well established. Scientific research results provoked a new term—sitting lifestyle death syndrome—whose indications are lower bone density, higher sugar levels in blood and urine, obesity, bad aerobic stamina, tachycardia in calm state, all of the mentioned being the source of disturbance for body organs and systems.

Unfortunately, as a result, today's generation of adolescents who are less than 25 years old, and make nearly half of the world's population, face far more complex challenges to their health and development than their parents did. It is thus important for people to become motivated regarding the importance of physical activity and nutrition for their health and wellbeing and be encouraged to practice a healthy lifestyle. Perception of healthcare should be based upon the fact that health starts with prevention, by carefully considering nutrition patterns, activity schedules and predisposition to several diseases, thus about our complete lifestyle.

### 3.2.2  Methodology for Qualitative Analysis of Human Body Surface Tissue Movement

Vertical movement is a major component in human daily physical activity. Walking, running, sitting up and down among them are the most common. Human body because of its properties can be treated as a skeleton with various tissues attached put into the "skin bag". Surface tissue, or "skin bag", is viscoelastic material that has different mechanical properties when deformed in direction perpendicular to and tangential to the surface. When vertical body movement is present mostly tangential surface deformation is present. Is it big enough even to be considered? Maybe human body surface tissue movement related to the skeleton is so small that it can be completely neglected?

A qualitative analysis can be a quick way to determine whether human body surface tissue can possibly induce errors big enough to be considered a problem. High speed camera might be used to capture vertical jump for example as such activity would exaggerate human body surface tissue vertical movement. Four dots can be drawn on the naked body to form an irregular shape. The size of the dots should be small enough—2 to 3 mm in diameter—and clearly visible. The size of the irregular shape should be big enough so the surface deformation could be captured (if it's present).

Some weight should be also attached to the same area to represent the acceleration measurement device so its impact could be taken into account as well. It's enough to make one jump to answer the questions that were raised earlier. The form of the irregular shape that is formed by given four dots can be compared in different jump stages. At least three stages should be analyzed: when human body is in steady state, when it's in upmost position and when it's in down most position. By comparing the shape of the defined form we can qualitatively conclude whether human body surface tissue deformations during vertical human body movement can be considered a problem.

### 3.2.3  Methodology for Quantitative Analysis of Human Body Surface Tissue Impact Towards Acceleration Measurements

If qualitative analysis of human body surface tissue movement during vertical jump shows that it may impact acceleration measurements, one would need to perform qualitative vertical jump analysis to identify how big that impact might be.

Taking into consideration that:

Acceleration measurement device is already developed and might be used;
Its attachment location is defined as chest area;

A methodology to quantitatively evaluate human body surface tissue impact towards acceleration measurements might be defined.

The goal of the experiment would be to measure movement of the skeleton of the human body during vertical jumping, to measure movement of the device that is attached to the chest area during the same jumping, and to compare acquired data by means of displacement and acceleration differences to conclude on the severity of the impact of human body surface tissue. The problem is, however, that measuring human body skeleton movement directly is impossible without intervention into the body itself. Thus the way to measure it indirectly must be defined. First, an area where soft tissue quantity on the bone is the lowest must be identified. It must be convenient to track and cannot restrict making a jump. Such area is the forehead because it satisfies all just mentioned requirements. Second, the way to remove all degrees of freedom from that area relative to the chest bone structure where the device is attached must be identified. The only loose area between the head and spine/ribs is neck. Thus, to achieve the required head and spine/ribs relation rigidity, neck ability to move the head must be completely removed so no nodding, head turning and everything in between would be possible. To achieve this, following measures can be used:

Neck support collar, similar to the one that is given in Fig. 3.9, tightly put on the subject's neck to remove its ability to move the head;
Rigid board with non-stretchable straps. The board would be put on the back with straps used to attach head, shoulders and pelvis so full torso rigidity could be achieved.

Test subject with neck collar, rigid board and acceleration measurement device attached is presented in Fig. 3.10.

With given setup in action full rigidity of head and spine/ribs is achieved. This way the forehead would completely resemble human spine/ribs movement during

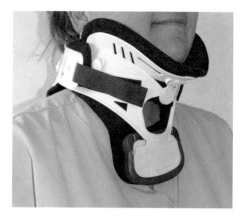

**Fig. 3.9**  Neck support collar [24]

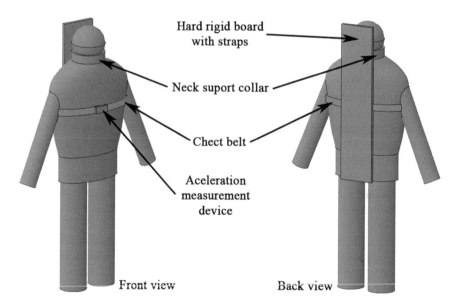

**Fig. 3.10**  Test subject with rigid board, neck support collar and acceleration measurement device attached

vertical jump. This setup, on the other hand, does not restrict surface tissue movement in the chest area.

Again, photogrammetry can be used to track the movement of the forehead and the device that is attached on the chest. Four markers should be attached for the cameras to track. One marker should be put on the forehead as a reference, and three more should be put on the acceleration measurement device to form two perpendicular virtual axes. The configuration of the markers is given in Fig. 3.11.

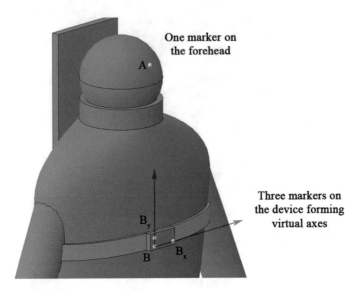

One marker on
the forehead

Three markers on
the device forming
virtual axes

**Fig. 3.11** Configuration of the markers

Marker on the forehead will be further referred as marker A, markers on the device will be further referred as B, $B_x$ and $B_y$.

Test subject should make multiple jumps with different heights in the described setup. Cameras would track markers A, B, $B_x$ and $B_y$ yielding 4 data sets. Each data set would then be filtered by using developed data filtering technique. Vertical displacement and acceleration data would be compared afterwards to conclude on the impact of human body surface soft tissue impact (how the device moves differently compared to the spine/ribs) towards the acceleration measurements.

Accelerometer measurement device was developed with digital acceleration signal bandwidth and amplitude characteristics in mind. However, the device must be validated and its operation analyzed before further use. Experiments to test developed acceleration measurement device (it will be simply referred as device) operation were conducted (Fig. 3.12).

Developed device was mounted on the vibration stand in such way that none of three accelerometer's axes would be aligned with vibration axis as required by the defined methodology (Fig. 3.13). Acceleration signal analysis revealed that the bandwidth of interest is 16 Hz. However, its worth expanding the bandwidth of interest up to 20 Hz for developed devices validation purposes. This will guarantee that no bad things might happen near the edge of bandwidth of interest (16 Hz). To analyze frequency range up to 20 Hz, 7 runs were made. Vibration stand was excited at the frequencies of 1, 4, 7, 10, 14, 17 and 20 Hz. Vibration stand movement was monitored while at least 10 oscillation periods were captured for each excitation frequency. An example of captured data is given in Fig. 3.14.

One should notice that vibration stand is able to produce oscillations of almost perfect sine wave as it is seen in Fig. 3.14. This shows that experimental equipment

**Fig. 3.12** Developed device validation experimental setup. *1* Frequencies generator Tabor Electronics WW5064 50 Ms/s; *2* Power amplifier VPA2100MN; *3* Vibration stand Veb Robotron Type 11077 with developed device mounted on top; *4* Displacement measurement unit laser Keyence LK-G82 with controller Keyence LK-GD500; *5* ADC Picoscope 3424

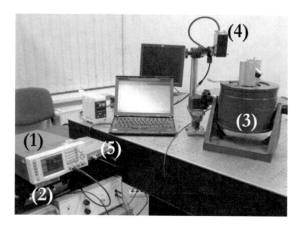

**Fig. 3.13** Device mounting on the vibration stand. *X*, *Y* and *Z* denotes global coordinate system; $X_D$, $Y_D$ and $Z_D$ accelerometer's coordinate system in the device

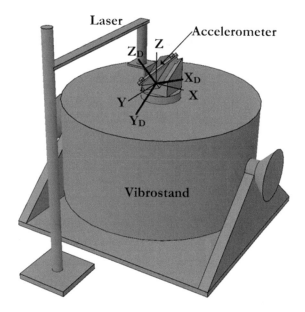

is adequately chosen and is up for a task. All collected vibration stand displacement data (seven data sets total) is subject for data filtering according to developed data filtering technique.

Frequency analysis was performed on each data set to identify frequency limit for the residual analysis. Frequency analysis results for the displacement data that was collected when vibration stand excitation frequency was set to 10 Hz, are given in Figs. 3.15 and 3.16.

Frequency analysis showed that the bandwidth of 100 Hz must be analyzed to identify safe pass band frequency for the low pass filter. Residual analysis plot of the displacement data that has been collected when vibration stand excitation

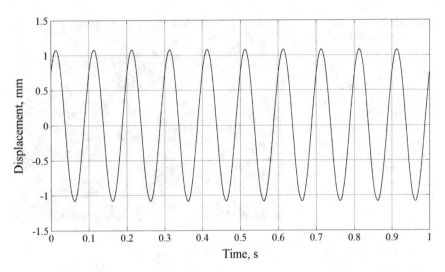

**Fig. 3.14** Sampled vibration stand displacement data for excitation frequency of 10 Hz

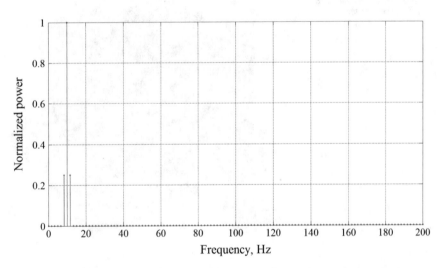

**Fig. 3.15** Normalized power plot of the displacement data that has been collected when vibration stand excitation frequency was 10 Hz

frequency was 10 Hz is given in Fig. 3.17. As can be seen in the plot, average residual increases linearly with decreasing filtering pass band frequency until the point when signal information starts to get lost. Then average residual starts increasing nonlinearly. In the given plot this point is around 11 Hz.

Complete residual analysis that was carried on all seven data sets showed that pass band should be set to 50 Hz (with a little reserve) to safely keep all signal information.

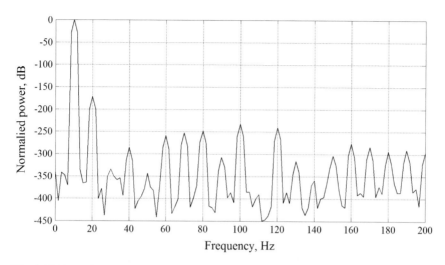

**Fig. 3.16** Normalized power (dB) plot of the displacement data that has been collected when vibration stand excitation frequency was 10 Hz

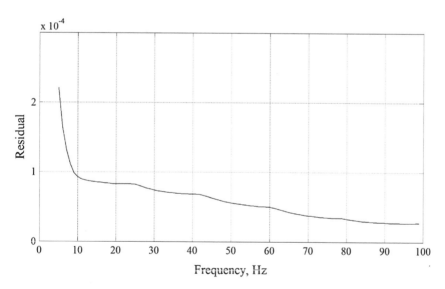

**Fig. 3.17** Residual analysis plot of the displacement data that has been collected when vibration stand excitation frequency was 10 Hz

Low pass filter was designed to have pass band of 50 Hz and full attenuation of 80 dB at 250 Hz (so the filter order would be acceptable). Filter characteristics are given in Table 3.7, magnitude response is given in Fig. 3.18 and phase response is given in Fig. 3.19.

**Table 3.7** Characteristics of designed 50 Hz low pass filter

| Characteristic | Value |
|---|---|
| Filter length | 167 coefficients |
| Pass band edge | 50 Hz |
| Stop band edge | 250 Hz |
| Stop band attenuation | 80.26 dB |
| Pass band ripple | 0.09 dB |

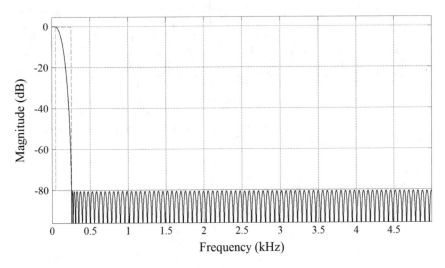

**Fig. 3.18** Magnitude response of designed 50 Hz low pass filter

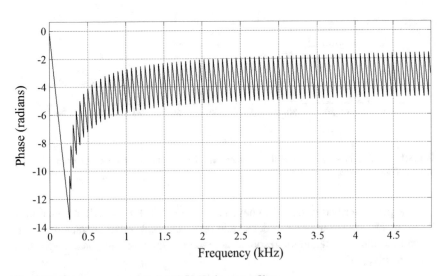

**Fig. 3.19** Phase response of designed 50 Hz low pass filter

**Table 3.8** Characteristics of the designed 30 Hz differentiator filter

| Characteristic | Value |
|---|---|
| Filter length | 123 coefficients |
| Pass band edge | 30 Hz |
| Stop band edge | 50 Hz |
| Stop band attenuation | 90.2 dB |
| Pass band ripple | 0.00016 dB |

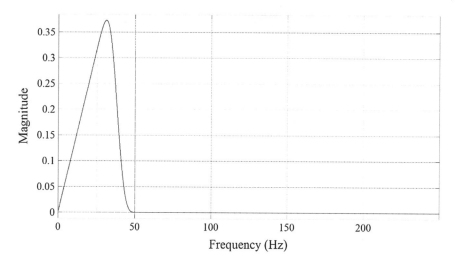

**Fig. 3.20** Magnitude response of the designed 30 Hz differentiator filter

After the data was lowpass filtered using designed lowpass filter, each data set was safely downsampled 20 times to final sampling frequency of 500 Hz (instead of 10 kHz) in order to lower computational load during analysis while still keeping undistorted signal. At this point all vibration stand displacement data sets are filtered and downsampled to 500 Hz. However, accelerations are needed as they will be compared to the output of the device. In order to acquire vibration stand movement accelerations 30 Hz differentiating filter was designed and applied twice on each set of data. Differentiating filter characteristics are given in Table 3.8, magnitude response is given in Fig. 3.20 and phase response is given in Fig. 3.21. Notice how magnitude and phase responses are linear in all required frequency bandwidth.

Designed differentiating filter was applied twice to obtain vibration stand movement accelerations which can be compared to the data from the device. Calculated vibration stand movement accelerations when vibration stand was excited at the frequency of 10 Hz is given in Fig. 3.22. Up scaled displacement curve is also given in the same plot for reference as a dotted curve.

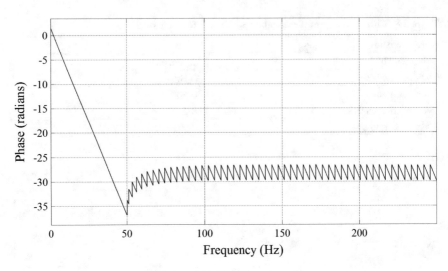

**Fig. 3.21** Phase response of the designed 30 Hz differentiator filter

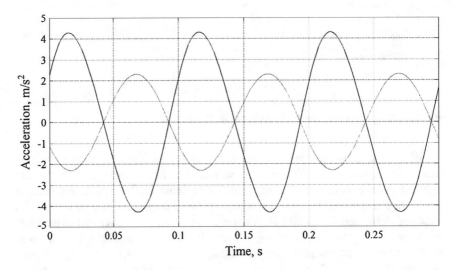

**Fig. 3.22** Calculated vibration stand movement accelerations when excitation frequency was set to 10 Hz

As described in the experiment setup, the device was mounted on top of the vibration stand. Acceleration measurements were sampled in parallel with vibration stand displacement data sampling for all used excitation frequencies. The device was set to sample acceleration data at the frequency of 400 Hz. An example of acceleration measurements that were collected by the device on each of accelerometer's axes when the vibration stand excitation frequency was set to

10 Hz, is given in Figs. 3.23, 3.24, and 3.25. Plotted accelerations already have the influence of acceleration due to removed gravity. That is why they oscillate around 0 but not around 9.8 m/s² (magnitude of vector of acceleration due to gravity) for all three axes.

Frequency analysis was performed on collected data samples. It suggested bandwidth of 50 Hz for residual analysis. After conducting residual analysis it was evident that pass band of 40 Hz must be set for lowpass filter. Thus lowpass filter was designed to have 40 Hz pass band frequency and 80 dB attenuation at frequency of 50 Hz. Then all data sets were filtered using designed low pass filter. An example of filtered acceleration data on each of accelerometer's axes when vibration stand excitation frequency was set to 10 Hz, is given in Figs. 3.26, 3.27, and 3.28. By comparing these three figures with the ones where unfiltered data is presented (Figs. 3.22, 3.23, 3.24, and 3.25), one can notice that high frequency noise that was present in unfiltered data and seen as curve disturbances in given plots, are removed after filtering resulting in smooth acceleration curves.

At this point vibration stand acceleration data and acceleration data that was acquired by the device are both filtered and ready for further steps. However, these two data sources cannot be yet compared to evaluate the operation of the device. Vibration stand accelerations are one dimensional data that lies on the vibration axis while accelerometer measurements by design are given as projections to accelerometer's axes and are three dimensional.

By employing least square method in 3D space the information on vibration axis is obtained from the acceleration data sets that were acquired with the device. Coefficient $k_x$, $k_y$, $k_z$, $b_x$, $b_y$, $b_z$ (Eq. 3.13) values that fully describe vibration axis were extracted from experimental data and are shown in Table 3.9.

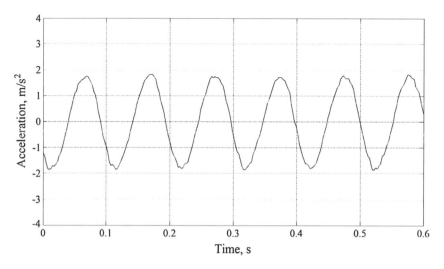

**Fig. 3.23**  Accelerations that were collected by the device on accelerometer X axis when vibration stand excitation was set to 10 Hz

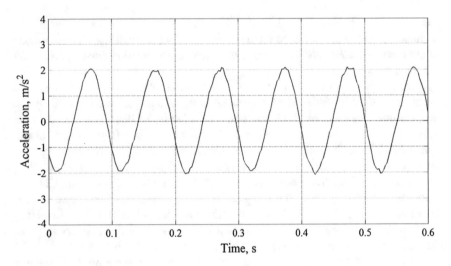

**Fig. 3.24** Accelerations that were collected by the device on accelerometer Y axis when vibration stand excitation was set to 10 Hz

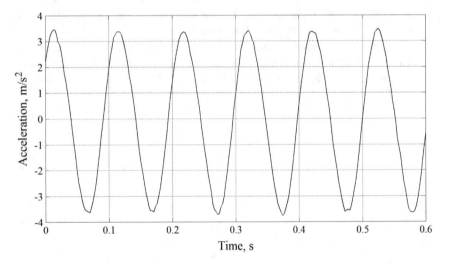

**Fig. 3.25** Accelerations that were collected by the device on accelerometer Z axis when vibration stand excitation was set to 10 Hz

An example of the acceleration data cloud plot with vibration axis identified when vibration stand excitation frequency was set to 10 Hz is given in Fig. 3.29.

Vibration stand position was not changed during the experiment. However, results in Table 3.9 show that vibration axis that was obtained from different data sets, is different. This situation is the consequence of all instrumental errors in the system. Still, data shows that with increasing excitation frequency the variance of

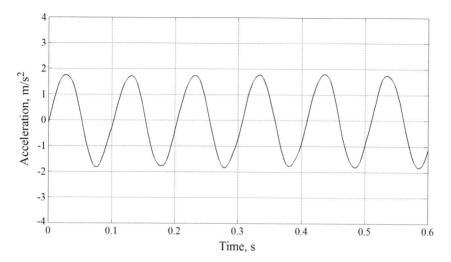

**Fig. 3.26** Filtered accelerations that were collected by the device on accelerometer X axis when vibration stand excitation was set to 10 Hz

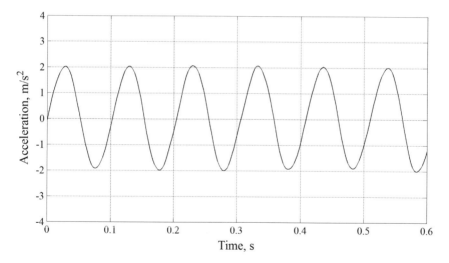

**Fig. 3.27** Filtered accelerations that were collected by the device on accelerometer Y axis when vibration stand excitation was set to 10 Hz

the data cloud in 3D space decreases. It is clearly seen by comparing the distribution of the data points in the cloud in Figs. 3.29 and 3.30.

Therefore, the axis of vibration, which has been obtained from the last set of data when vibration stand excitation frequency was set to 20 Hz is selected as the true axis. Furthermore, the least square method resulted with the lowest error compared to other data sets. Coefficients for that axis are given in Table 3.10.

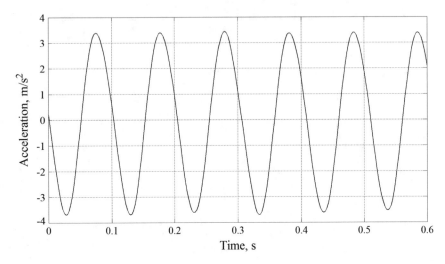

**Fig. 3.28** Filtered accelerations that were collected by the device on accelerometer Y axis when vibration stand excitation was set to 10 Hz

**Table 3.9** Coefficient values that fully describe vibration axis for each excitation frequency

| Excitation frequency, Hz | $k_x$ | $k_y$ | $k_z$ | $b_x$ | $b_y$ | $b_z$ |
|---|---|---|---|---|---|---|
| 1 | 0.5310 | 0.4330 | −0.7284 | 0.0001 | 0.0228 | 0.0037 |
| 4 | 0.4016 | 0.4639 | −0.7897 | 0.0002 | 0.0182 | 0.0197 |
| 7 | 0.4155 | 0.4491 | −0.7910 | 0.0056 | 0.0306 | −0.0048 |
| 10 | 0.4072 | 0.4514 | −0.7940 | 0.0114 | 0.0449 | −0.0674 |
| 14 | 0.4062 | 0.4447 | −0.7983 | −0.0040 | 0.0242 | −0.0579 |
| 17 | 0.4039 | 0.4360 | −0.8042 | −0.0212 | 0.0072 | −0.0461 |
| 20 | 0.4068 | 0.4259 | −0.8081 | −0.0176 | 0.0265 | −0.0708 |

After obtaining vibration axis information, all acceleration data that was captured with the device, was transformed in space so accelerometer's Z axis would be aligned with obtained vibration axis. This alignment is required so data would have the same reference and could be compared. Again, MATLAB routines were used for this task. An example of in a described way aligned data for every accelerometer axis is given in Figs. 3.31, 3.32, and 3.33. Excitation frequency of vibration stand was 10 Hz.

It is important to state that after described alignment accelerations on accelerometer's X and Y axes became practically zero which was the desired outcome. Small accelerations are still present but they are explained by accelerometer cross axis sensitivity, which means that acceleration on one axis can be "seen" on other axis although with much lower magnitude.

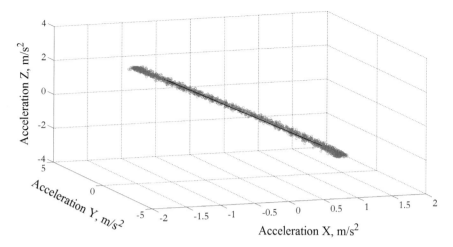

**Fig. 3.29**  Acceleration data cloud with identified vibration axis when vibration stand excitation frequency was set to 10 Hz

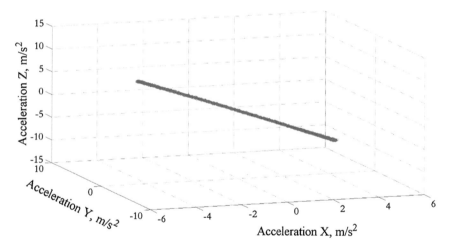

**Fig. 3.30**  Acceleration data cloud with identified vibration axis when vibration stand excitation frequency was set to 20 Hz

**Table 3.10**  Final vibration axis coefficient values

| $k_x$ | $k_y$ | $k_z$ | $b_x$ | $b_y$ | $b_z$ |
|-------|-------|-------|-------|-------|-------|
| 0.4068 | 0.4259 | −0.8081 | −0.0176 | 0.0265 | −0.0708 |

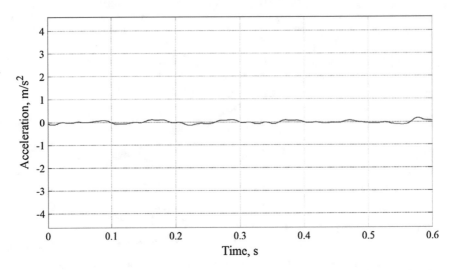

**Fig. 3.31** Accelerations that were collected by the device on accelerometer X axis when vibration stand excitation was set to 10 Hz and accelerometer's Z axis was aligned with vibration axis

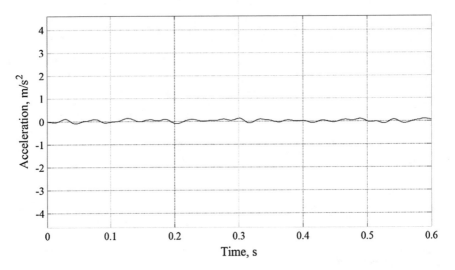

**Fig. 3.32** Accelerations that were collected by the device on accelerometer Y axis when vibration stand excitation was set to 10 Hz and accelerometer's Z axis was aligned with vibration axis

Now it is possible to compare designed device measurements with vibration stand accelerations to conclude on the device applicability in further research. Table 3.11 summarizes comparison results.

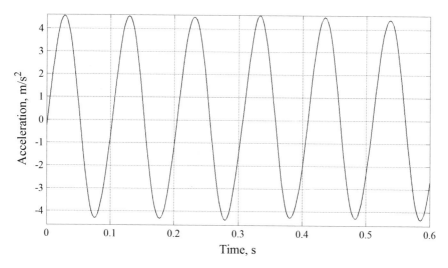

**Fig. 3.33** Accelerations that were collected by the device on accelerometer Z axis when vibration stand excitation was set to 10 Hz and accelerometer's Z axis was aligned with vibration axis

**Table 3.11** Acceleration measurement errors for each excitation frequency

| Frequency, Hz | Excitation amplitude, mm | Acceleration amplitude, m/s² | Measured acceleration amplitude, m/s² | Error, % |
|---|---|---|---|---|
| 1 | 0.8813 | 0.3480 | 0.4018 | 15.46 |
| 4 | 0.9517 | 0.6772 | 0.7668 | 13.22 |
| 7 | 0.9510 | 1.8397 | 2.0435 | 11.07 |
| 10 | 1.0822 | 4.2725 | 4.5236 | 5.88 |
| 14 | 1.0335 | 7.9967 | 8.1595 | 2.04 |
| 17 | 0.9891 | 11.2849 | 11.0833 | 1.79 |
| 20 | 0.9241 | 14.5920 | 13.8016 | 5.42 |

Experimental results show that the measurement error for frequencies over 7 Hz gives an error of 11%. Lower frequencies tend to produce large errors because of low values of acceleration, so the device operates near its signal-to-noise ratio limit. It must be said that the actual movement of the different places on the human body has a much greater amplitude in the range of tens of centimeters. This means that low frequency movements still result in higher levels of acceleration, and further the device can operate on its signal to noise ratio. These levels are a way to reduce errors to acceptable limits in real-life measurements even with a frequency of up to 7 Hz.

### 3.2.4 Qualitative Analysis of Human Body Surface Tissue Movement During Vertical Jump

By following the methodology that is given in Sect. 3.2.3, qualitative vertical jump assessment was made during the experiment. Video was captured by Phantom ir 300 high speed camera that is capable to deliver up to 100,000 fps. During vertical jump video capture the camera was set to 4000 fps at the resolution of 800 × 600 pixels. 2 m distance was set between the camera and the subject. Two 500 W halogen spotlights were set up for adequate lightning for the camera to operate efficiently. Video of the jump was digitally recorded into the camera memory, downloaded to personal computer and analyzed using video player with more sophisticated features like ability to set play speed and to capture video frames into the images.

A small end user device (weight of approx 50 g) was attached to the belly using leather belt (not stretchable). Four dots were painted on the skin (Fig. 3.34) to form irregular shape as reference. The size of the shape fits into the area of 8 × 15 cm. Initial steady state position is given in Fig. 3.35.

Video of the jump was captured using super high frame rate camera. Irregular shapes at two extremities—upmost body position (Fig. 3.35, left) and down most position (Fig. 3.35, right) during jump—were compared with a shape in a steady state.

These three frames were put together (Fig. 3.36) with one top of each shape aligned to see actual differences between the shapes during various vertical jump phases. As expected, soft tissue stretched during take-off phase as lower points were a little bit 'behind' the top ones. Soft tissue compressed during hard landing as the top points were still moving down while the bottom one already reached it's down most position.

One of the side's length was measured in steady state and was 15 cm. The same side was approximately measured with indirect tool—images editing software. In the upmost position side's length was ∼17.7 cm while in down most position it was ∼13.9 cm. This equals to 18% length change in upmost position and −7.3% length change in down most position.

**Fig. 3.34** Four dots forming irregular shape for tracking. Image shows a steady state with end user device attached

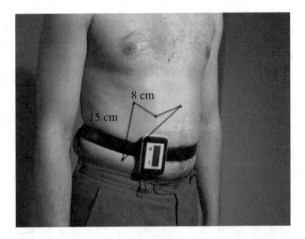

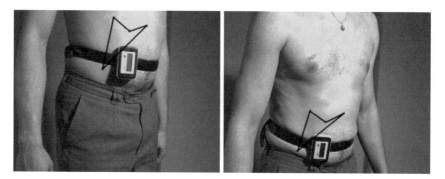

**Fig. 3.35** Two jump process extremities: *left* upmost position; *right* down most position

**Fig. 3.36** Comparison of formed irregular shapes during vertical jump: *1* steady state; *2* upmost jump position; *3* down most position

This qualitative assessment shows that skin surface moves differently during different vertical jump stages. Relative shape side length change was observed to reach approximately 18% when compared with a steady state. This means that surface tissue movement during vertical human body motion cannot be neglected and the impact towards acceleration measurements must be quantitatively evaluated.

### 3.2.5   Quantitative Analysis of Human Body Surface Tissue Impact Towards Acceleration Measurements

In the previous section it was concluded that human body surface tissue movement cannot be neglected during vertical human body movement. Thus, qualitative analysis of the impact for the acceleration measurements is needed. To analyze the size of the impact, a number of jumps were recorded with a described experimental setup using sampling frequency of 100 Hz. Data was collected by tracking four points A, B, $B_x$ and $B_y$ (Fig. 3.11). Measurements reference is given in Fig. 3.37. An example of data that was captured by tracking marker A is given in Figs. 3.38, 3.39, and 3.40.

All acquired data sets were filtered using designed filtering technique. Frequency analysis was performed on collected data and showed that 20 Hz bandwidth is sufficient for further analysis as normalized logarithmic power spectra plots showed

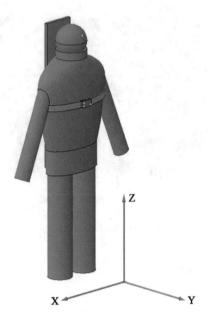

**Fig. 3.37** Measurements reference coordinate system

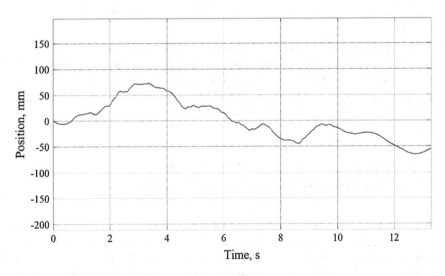

**Fig. 3.38** X coordinates acquired by tracking marker A

stable signal level over 20 Hz. Residual analysis was performed and suggested a pass band frequency of 8 Hz. Lowpass filter was constructed with pass band of 8 Hz as provided by signal residual analysis, stop band at 10 Hz and stop band attenuation of 80 dB. Filter characteristics are given in Table 3.12.

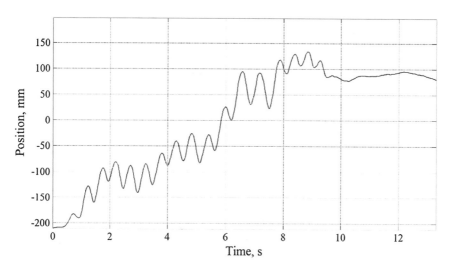

**Fig. 3.39** Y coordinates acquired by tracking marker A

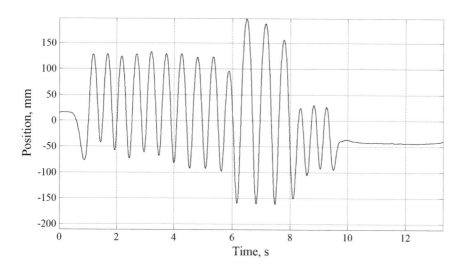

**Fig. 3.40** Z coordinates acquired by tracking marker A

In order to quantitatively assess vertical jumping three distances variations in time were analyzed throughout the jumping process:

Distance between markers A and B;
Distance between markers B and $B_x$;
Distance between markers B and $B_y$.

  Refering to Fig. 3.11 for the configuration of the markers.
  All three distance variations in time are given in Fig. 3.41.

**Table 3.12** Characteristics of designed 8 Hz lowpass filter

| Characteristic | Value |
|---|---|
| Filter length | 181 coefficients |
| Pass band edge | 8 Hz |
| Stop band edge | 10 Hz |
| Stop band attenuation | 79.95 dB |

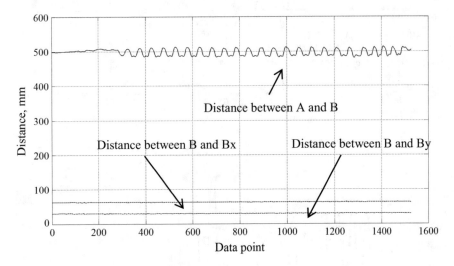

**Fig. 3.41**  Three distances variation in time

As expected distances between markers on the device (pairs B and $B_x$, B and $B_y$) were stable as the device is rigid body (obviously). However, distance between markers A and B varied from jump to jump. A direct comparison of jump height that was measured by tracking marker A and height that was measured by tracking marker B is given in Table 3.13 and Fig. 3.42.

Average jump height that was measured by tracking marker A was 137 mm, while jump height that was measured by tracking marker B was 154 mm, giving average difference of 17 mm or 11% higher values. Jumping was periodic signal with average frequency of 1.9 Hz.

Close comparison between Z axis data for markers A and B is given in Fig. 3.43. One can clearly see how the jumping curve is different because of human body surface tissue rheology. Forehead point (marker A, or the point that corresponds to bone fixed point movement) has lower oscillation amplitude than the device (marker B, corresponds to the point whose movement is dependent on soft tissue properties), although the frequency and the phase are the same because of the low movement frequency.

Acceleration on Z axis for marker B was 34% higher on average (Fig. 3.42) compared to acceleration of the marker A for the accelerations up to 27 m/s$^2$.

**Table 3.13** Vertical jumping data

| Jump height measured by tracking marker A, mm | Jump height measured by tracking marker B, mm | Difference, mm |
|---|---|---|
| 127 | 143 | 16 |
| 134 | 151 | 17 |
| 133 | 150 | 17 |
| 136 | 152 | 16 |
| 140 | 157 | 17 |
| 140 | 159 | 19 |
| 147 | 166 | 19 |
| 144 | 161 | 17 |
| 144 | 161 | 17 |
| 128 | 143 | 15 |

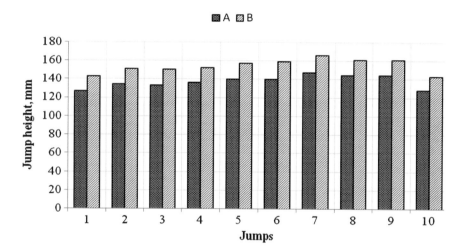

**Fig. 3.42** Vertical jump heights measured by tracking markers A and B

This gives rise to a very important conclusion: because of the surface of the body tissue rheological properties of the human chest device movement relative to the spine/ribs (according to described methodology marker A corresponds to that) movement is not identical and human body surface tissue rheological properties strongly impacts acceleration measurements (Fig. 3.44).

Qualitative vertical jump assessment which was performed on vertical jumping proved that human body surface tissue is the source of errors when analyzing human body motion. Collected data was the subject for frequency and residual analysis which showed that jumping acceleration data falls into bandwidth of 8 Hz. Vertical jumping data analysis also revealed that average jump height measured by forehead point (marker A) was 137 mm, while jump height measured by point on

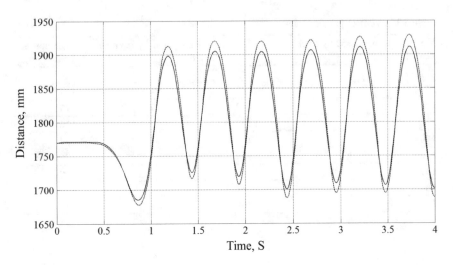

**Fig. 3.43** Z distances of markers A (*solid line*) and B (*dotted line*)

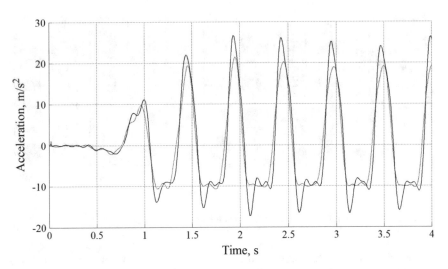

**Fig. 3.44** Acceleration on Z axis differences between markers A (*dotted line*) and B (*solid line*)

the device (marker B) was only 154 mm, giving average difference of 17 mm or
11% higher values. Acceleration values with human body surface tissue present and
without it were compared. It was observed that acceleration on Z axis of the device
was 34% higher on average compared to acceleration of the forehead point for the
accelerations up to 27 m/s$^2$.

### 3.2.6 Multi-level Computational Model

The main goal of this work is the introduction of a tool that can be applied directly to the surface of the process which is measured and by the acceleration of the human body tissue induced errors in this way can be reduced. The surface of the human body tissues induced inaccuracies, as it was presented in 3.2.3, can increase the value of the measured accelerations up to 35% when the acceleration value of up to 27 m/s$^2$ is measured. Although the time frames of these interruptions are short—up to 0.2 s—they still lead to an increase in the total measurement error. In order to address this issue, a computational model must be developed. This model would serve as a tool to model and predict resulting errors, and as a result, reduce these errors up to a tangible margin.

The area of chest with the acceleration measurement device attached by the belt is the primary area of interest. Let's take the experimental setup that is described in Sect. 3.2.3. Closer look is given in Fig. 3.45.

First step would be to develop a reduced human body surface tissue rheological model that corresponds to the area A that is given in Fig. 3.45. A developed reduced model can be introduced with vertical jump data that was collected during quantitative analysis of human body surface tissue impact towards acceleration measurements. That data would serve as the excitation law for the model, and the movement of the device that is attached on the chest would be observed. By comparing that output with the device movement data that was captured with cameras and markers during the same experiment developed model could be validated.

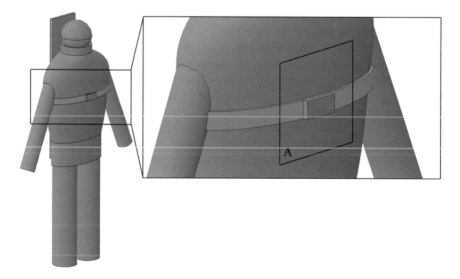

**Fig. 3.45** Area of interest (*A*) for the reduced human surface tissue rheological model

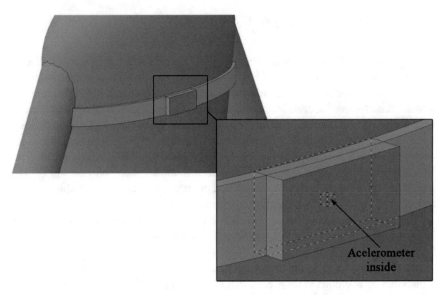

**Fig. 3.46** Accelerometer that is present inside acceleration measurement device

Once the model is validated and corresponds to the experiment data, it is possible to go deeper into the acceleration measurement device and develop an accelerometer model that would yield the same output as accelerometer would (Fig. 3.46).

Developed accelerometer model could be validated by using the data that was collected during the validation of the developed acceleration measurement device operation. Vibration stand movement data could be used as an excitation law for the model, and the output of the model would be compared with the actual measurements that were taken by the device itself.

Both of mentioned models are part of one system and thus should be analyzed as such. For this reason it is important to determine whether those models can be combined into complex multi-level computational model that would allow modeling the accelerometer's output based on external excitation of the whole system. The scheme of such approach is given in Fig. 3.47.

Developed multi-level computational model then would serve as the main tool to lay the base for the error reduction algorithm.

### 3.2.7  Reduced Human Body Surface Tissue Rheological Model and Its Validation

Human body is a complex system that is made up from very different materials with very different purposes ant properties. All cardiovascular system; bio viscoelastic fluids; skeletal, heart and smooth muscles; bones and cartilages are among them and

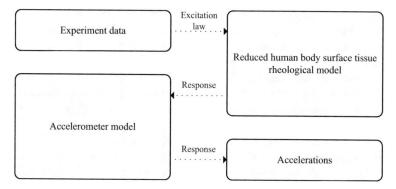

**Fig. 3.47** Reduced human body surface tissue rheological and accelerometer models combined into one multi-level computational model

represent huge fields of biomechanics by themselves. These topics can be further split and analyzed until full human body decomposition is achieved. Usual (common) approach, when these materials are analyzed, is divided into the steps given below:

1. Study of the morphology of the organism, organ anatomy, histology tissue as well as the structure and ultrastructure of the material in order to know the geometric configuration of the object of interest.
2. Determine the mechanical properties of a material or tissue that are involved in the problem. In biomechanics, this step is often very difficult, either because it is impossible to isolate the tissue for testing, or because the amount of available tissue samples is too small, or because it is difficult to keep the tissue in a normal living condition. In addition, the soft tissue is often exposed to great strain and the stress-strain relationship is usually nonlinear and depends on history.
3. Obtain governing differential or integral equations on the basis of the fundamental laws of physics and constitutive material equations, understand the environment in which the body works to get meaningful boundary conditions.
4. Solve the boundary value problems analytically or numerically, or by means of experiments.
5. Follow the physiological experiments, which will be tested for solving boundary value problems mentioned above.
6. Compare the experimental results with the corresponding theoretical knowledge. Through comparison, it is determined whether the hypothesis made are justified in theory, and if they have, find the numerical values of undetermined coefficients in constitutive equations.
7. Use the confirmed theory to predict the outcome of the other boundary value problems connected with the same basic equations. Then the method to study the practical application of the theory and experiments could be used.

Steps #1 and #2 were left aside because the objective is to develop a reduced human body surface tissue rheological model that would have properties not related to the specific soft tissue or material but to the combination of all related materials in area of interest (see Fig. 3.45). This is the only approach possible when analyzing human body or its part as a complex system as a combination of complex subsystems which are also a combination of other systems or materials and so on. The most serious disorder usually lack of information about living tissue the constitutive equations. Without the foundation of laws, no analysis can not be done. Thus, one finds oneself in a situation in which serious analyzes (as a rule, is difficult because of the nonlinearity) should be made for the hypothetical material, in the hope that the experiments will yield the desired agreement.

When the problem is to determine body surface movement that is the outcome of "chain reaction" of several different biomechanical materials and their interaction (bones, different types of muscles, internal organs, blood and other fluids as well as fat and skin), it is impractical at least, and impossible at most to follow the path mentioned above.

That is why a reduced human body surface tissue rheological model is proposed [25] to be developed so practical application of surface soft tissue artifacts reduction can be explored. The chest area, that is the area of interest (Fig. 3.45), is "cut" from the body to form the model that consists of two different parts:

1. An acceleration measurement device with initial pre-stress which corresponds to the belt that is used to attach the device;
2. Reduced human body surface tissue rheological model where the acceleration measurement device is attached.

Model scheme is presented in Fig. 3.48.

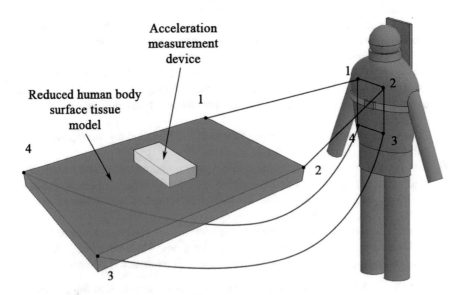

**Fig. 3.48** Model scheme showing its location and parts

The acceleration measurement device (Fig. 3.48) with micro accelerometer designed in previous subchapter is selected to be isotropic elastic material. Its physical geometrical size is $80 \times 40 \times 16$ mm. Its mass is 42 g. This way its density can be said to be 806 kg/m$^3$.

Initial pre-stress is described as:

$$\mathbf{K} \cdot \mathbf{u} = \mathbf{F} \tag{3.22}$$

where K is stiffness matrix, u is displacement vector and F is force vector.

Reduced human body surface tissue rheological model (Fig. 3.8) is selected to be hyper elastic Neo-Hookean material. Its geometrical size is selected $300 \times 200 \times 20$ mm. Given size completely covers the area where the acceleration measurement device movement mismatch exists. Initial shear modulus is chosen to be 100 Pa [26], and initial bulk modulus is chosen to be 0.1 GPa [27, 28]. Density was set to 1000 kg/m$^3$ that is the density of the water (human body is over 2/3 of water).

Given parameters were used to make a FE model in Comsol multiphysics. Meshed model is given in Fig. 3.49. Contact between acceleration measurement device and reduced human body surface tissue rheological model is taken into account.

This finite element formulation of the dynamics of the reduced model of the human body is described by the following equation of motion in the form of a block, taking into account that the law of motion of the base is known and determined the nodal displacement vector uK(t):

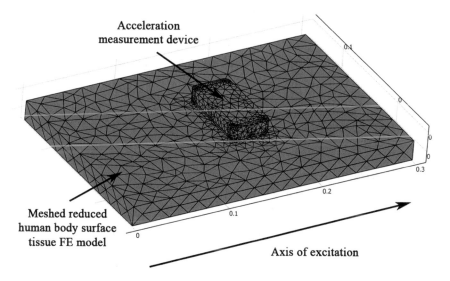

**Fig. 3.49** Meshed FE model in the COMSOL with excitation axis presented

$$\begin{bmatrix} \mathbf{M}_{NN} & \mathbf{M}_{NK} \\ \mathbf{M}_{KN} & \mathbf{M}_{KK} \end{bmatrix} \cdot \begin{bmatrix} \mathbf{u}''_N \\ \mathbf{u}''_K \end{bmatrix} + \begin{bmatrix} \mathbf{C}_{NN} & \mathbf{C}_{NK} \\ \mathbf{C}_{KN} & \mathbf{C}_{KK} \end{bmatrix} \cdot \begin{bmatrix} \mathbf{u}'_N \\ \mathbf{u}'_K \end{bmatrix}$$
$$+ \begin{bmatrix} \mathbf{K}_{NN} & \mathbf{K}_{NK} \\ \mathbf{K}_{KN} & \mathbf{K}_{KK} \end{bmatrix} \cdot \begin{bmatrix} \mathbf{u}_N \\ \mathbf{u}_K \end{bmatrix} = \begin{bmatrix} \mathbf{0} \\ \mathbf{r} \end{bmatrix} \tag{3.23}$$

where nodal displacement vectors uN(t) and uK(t) correspond to displacement of free and kinematically excited nodes, respectively; M, C and K mass, damping and stiffness matrices, respectively; r is a vector representing the reaction force kinematically excited nodes.

Displacement vector of unconstrained nodes is expressed as:

$$\mathbf{u}_N = \mathbf{u}_{Nrel} + \mathbf{u}_{Nk} \tag{3.24}$$

where $u_{Nrel}$ denotes a component of relative displacement with respect to moving base displacement $u_{Nk}$.

Vectors $u_{Nk}$ and $u_K$ define corresponding movement of a rigid body, which do not cause internal elastic forces in the structure. Proportional damping approach is adopted in the form of:

$$\mathbf{C} = \alpha \cdot \mathbf{M} + \beta \cdot \mathbf{K} \tag{3.25}$$

with $\alpha$ and $\beta$ as Rayleigh damping constants.

Consequently, the following matrix equation is obtained after algebraic rearrangements of previous equations and contains a matrix structure in the constrained nodes imposed by kinematic excitation:

$$\mathbf{M}_{NN} \cdot \mathbf{u}''_{Nrel} + \mathbf{C}_{NN} \cdot \mathbf{u}'_{Nrel} + \mathbf{K}_{NN} \cdot \mathbf{u}_{Nrel} = \widehat{\mathbf{M}} \tag{3.26}$$

Here $\widehat{\mathbf{M}}$ represents a vector of inertial forces that act on each node of the structure as a result of applied kinematic excitation and is expressed as:

$$\widehat{\mathbf{M}} = \mathbf{M}_{NN} \cdot \mathbf{K}_{NN}^{-1} \cdot \mathbf{K}_{NK} - \mathbf{M}_{NK}. \tag{3.27}$$

The kinematic excitation was imposed on the boundary in terms of displacement vector $u_K(t)$.

Model validation is performed by utilizing experimental vertical jump data that was acquired during quantitative analysis of human body surface tissue impact towards acceleration measurements. Model validation is achieved by doing the following:

1. The vertical movement of the marker A (Fig. 3.11) that corresponds to the vertical movement when no surface tissue impact is presented is used as an excitation law for the reduced human body surface tissue rheological model.

2. The response of the model is observed in COMSOL environment by monitoring the movement of the acceleration measurement device that is attached (Fig. 3.49). Resulting device movement is extracted into separate data files.
3. Modeled movement of the acceleration measurement device is compared to the experimental device movement data to see how well the model fits the experiment.
4. Conclusions are drawn based on the comparison results.

Experimental vertical jump data that corresponds to the movement when no surface tissue impact is present was used as an excitation law and is given in Fig. 3.50. Experimental movement of the attached acceleration measurement device is also given in Fig. 3.50.

To compare the output of the model with experimental data all curves where plotted in one figure (Fig. 3.51). Three curves are given. Solid curve represents the vertical movement of the underlying spine/ribs entity when no surface tissue impact is present. Dotted line represents the experimentally acquired acceleration measurement device movement. The device was attached to the chest with a belt and represents the movement with soft tissue impact present. Finally, dashed curve represents modeled movement of the acceleration measurement device. The modeled output was acquired from the developed reduced human body surface tissue model when the data that the solid curve represents was used as an excitation law.

As can be seen in Fig. 3.51 modeling results fit the experiment data. However, one fitment is not enough to validate that the model corresponds to the real world object. Figure 3.52 shows that modeling results fit also another experiment data. Three curves are given in the same manner as in Fig. 3.51. Solid curve represents the vertical movement of the underlying spine/ribs entity when no surface tissue

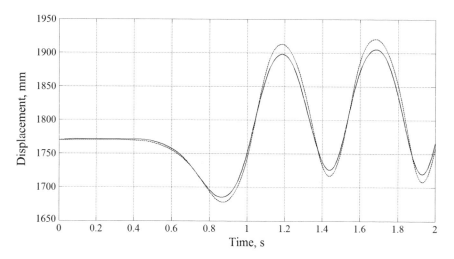

**Fig. 3.50** Experimental vertical jump data (*solid line*) and movement of the acceleration measurement device (*dashed line*)

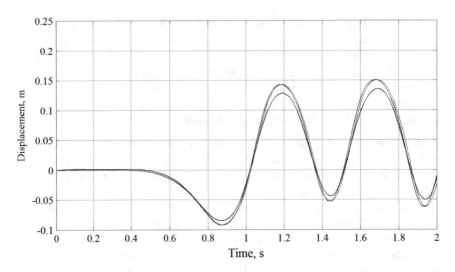

**Fig. 3.51** Experimental vertical jump data when no surface tissue impact is present (*solid line*); acceleration measurement device movement during experiment (*dotted line*); modeled acceleration measurement device movement (*dashed line*)

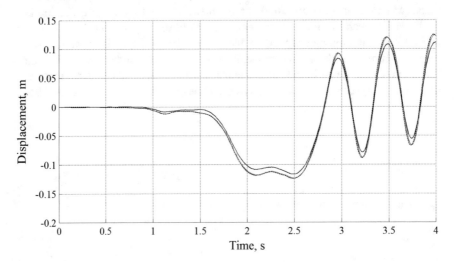

**Fig. 3.52** Experimental vertical jump data when no surface tissue impact is present (*solid line*); acceleration measurement device movement during experiment (*dotted line*); modeled acceleration measurement device movement (*dashed line*)

impact is present. Dotted line represents the experimentally acquired acceleration measurement device movement. Finally, dashed curve represents modeled movement of the acceleration measurement device.

Average error is defined as:

$$e = \frac{\sum_{i=1}^{N} |y_i - \widehat{y}_i|}{N} \tag{3.28}$$

where $y_i$ is experiment data point; $\widehat{y}_i$ is modeled data point; N is number of data points.

For the first experiment data set average error was 0.0034, maximum error was 0.0136; for the second experiment data set average was 0.0023, maximum error was 0.0132.

This means that the average difference between experiment data point and model output data point was 0.0034 m (3.4 mm) for the first experiment and 0.0023 m (2.3 mm) for the second experiment.

Average jump height measured by tracking marker A (Fig. 3.11), was 154.3 mm (Table 3.13). If this average is taken as a base, it can be said that model has average relative error of 2.2% for the first experiment data and relative error of 1.5% for the second experiment data. Maximum relative error for both experiments was lower than 8.9%.

It can be concluded that the model corresponds well to the real world object – the model was developed using one experimental data, and successfully matched another experimental data with low errors.

## 3.2.8   Reduced Human Body Surface Tissue Rheological Model Analysis

Reduced human body surface tissue rheological model was developed and validated in the previous section. Validated model is further analyzed to get deeper insight in its behavior. Axes of reference used during the analysis are given in Fig. 3.53. Axis X corresponds to the vertical body movement.

Figure 3.54 shows how the displacement on the skin surface is different through the area reaching its maximums just under the attached device. During jump whole soft tissue is moving differently from underlying bones, however attached device impacts surrounding area and introduces additional tissue displacement that increases gradually in magnitude when going towards the device.

As the device is a 3D body it not only affects the surrounding soft tissue area along jump axis (axis X) but also introduces small displacement on axis that is perpendicular to the jump axis, as well as surface traction forces around the device. This is also expected behavior because during jump the device's top tends to flip off the body while pressing the area at the device bottom. This effect is clearly visualized in Figs. 3.55 and 3.56.

Interesting behavior is observed when comparing two stages of the jump: the upper position when the speed is 0 and speed maximum during landing. At the most

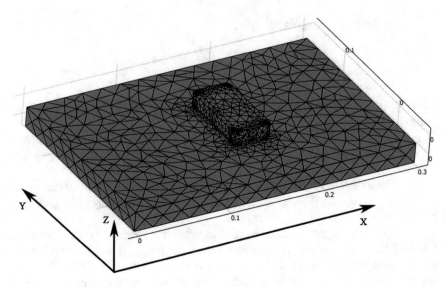

**Fig. 3.53** Axes of reference used during analysis of the model

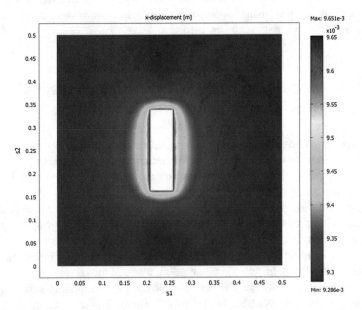

**Fig. 3.54** Reduced human body surface tissue rheological model displacement on X axis

upper point when the body has already stopped going up the attached device still
has kinetic energy and tries to go up further. As it is attached to the body it cannot
do so thus it creates soft tissue surface traction forces around it. This situation is

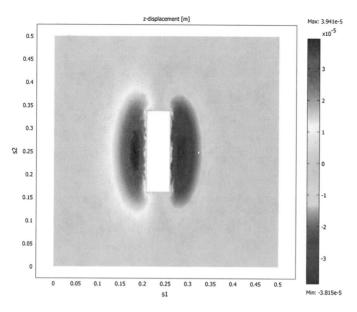

**Fig. 3.55** Reduced human body surface tissue rheological model displacement on Z axis

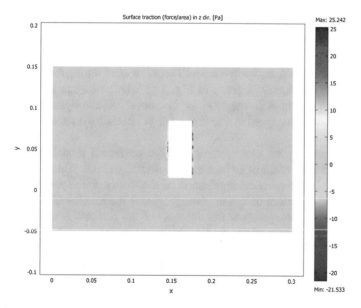

**Fig. 3.56** Reduced human body surface tissue rheological model surface traction

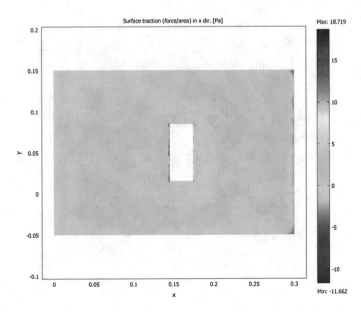

**Fig. 3.57** Reduced human body surface tissue rheological model surface traction in X direction when the speed is 0

visualized in Figs. 3.57 and 3.58. On the contrary, when the body is landing and the movement speed is at its maximum, surface traction forces are only induced by the mass of the device. This is visualized in Figs. 3.59 and 3.60.

It is very interesting to observe model's cross sections during the process. One of such cross section figure is given in Fig. 3.61 where one can see how the displacement changes in different depth of the model. As the figure is given in the jump phase when the speed is the highest it is interesting to see the response of non-linear models.

As mentioned in the previous section, the confirmed model can be used to predict the outcome of the other problems associated with the same model. Then the method to study the practical application of the theory and experiments could be used. One such practical task is to identify what is the relationship between initial pre-stress which corresponds to the tightness of the belt that is used to attach the device and the response of the model which corresponds to the movement of the device. Such relationship would describe how the attachment belt tightening impacts device movement (thus, soft tissue movement as well) on the chest during vertical movement of the body (that is the most common movement during daily activities).

Reduced human body surface tissue rheological model was used to acquire the applied pressure needed to reach different device's displacement into the body. The relationship between the displacement and the pressure applied is given in Fig. 3.62. As can be seen in given figure, the relationship is almost linear in the

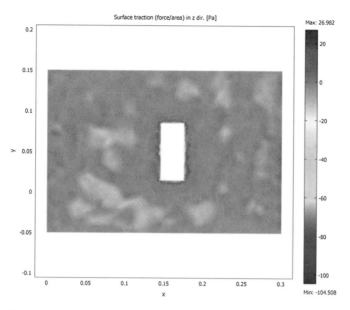

**Fig. 3.58** Reduced human body surface tissue rheological model surface traction in Z direction when the speed is 0

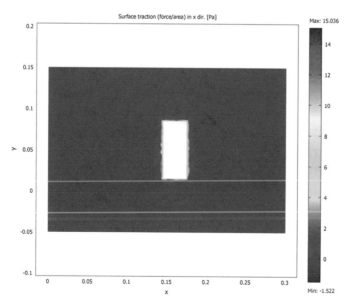

**Fig. 3.59** Reduced human body surface tissue rheological model surface traction in X direction when the speed is maximum during landing

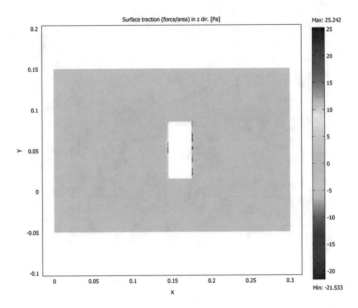

**Fig. 3.60** Reduced human body surface tissue rheological model surface traction in Z direction when the speed is maximum during landing

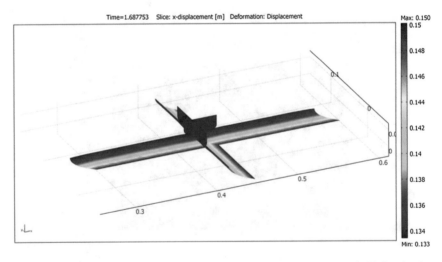

**Fig. 3.61** Reduced human body surface tissue rheological model displacement in X direction just before take-off cross section

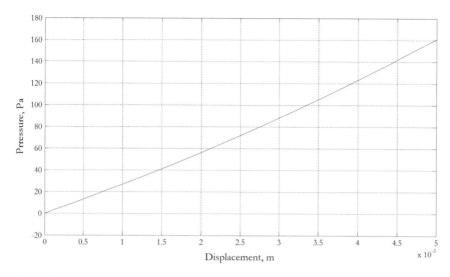

**Fig. 3.62** On the device applied pressure relationship with its displacement into soft tissue

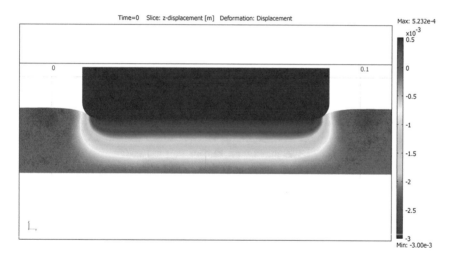

**Fig. 3.63** Reduced human body surface tissue rheological model displacement on axis Z when the device is pushed 3 mm into the body

displacement range of 0 to 5 mm. An example of the displacement variations inside the model is given in Fig. 3.63.

To acquire model's response dependence on initial pre-stress (pressure applied to the device to push it into the surface tissue), four different device's displacements into the body were analyzed: 0, 1, 2 and 3 mm.

Developed reduced human surface tissue was provided with different displacements of the device into the body. Then, the model was excited using the

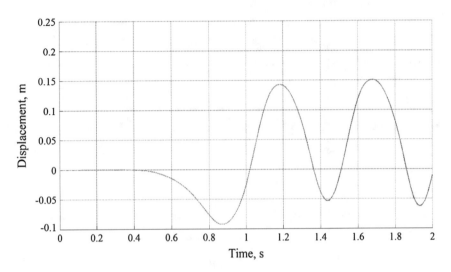

**Fig. 3.64** Combined plot of modeled device movements when different device displacement into the body were set

experimental data that was acquired during quantitative analysis of human body surface tissue impact towards acceleration measurements. Finally the modeled movement of the device was observed and compared in between when different device's initial displacements into the body were used. The plot off all outputs is given in Fig. 3.64.

Modeling results show that reduced human body surface tissue rheological model response does not depend on initial device displacement into the body. This means that the belt tightening force that is used to attach the device is not a factor for the vertical surface tissue movement during vertical movement of underlying bones. Although it might seem strange at first, it is logical conclusion as the belt moves together with tissue surface (skin) and thus its movement depends only on the properties of the tissue rather than the tightening force. This conclusion is important in practice too allowing neglecting belt tightening as a factor.

Another two factors of interest are jumping height and jumping frequency. From physics it is known that vertical movement of point obeys this equation:

$$s = v_0 \cdot t + \frac{g \cdot t^2}{2} \tag{3.29}$$

where s is path travelled, $v_0$ is initial velocity, g is the acceleration due to gravity, and t is time.

In a vertical jump, half of the path travelled is up and half is down. In the upper jump point velocity of the spine/ribs is 0. If we take upper point as initial process condition than the process of landing can be defined as:

$$h = \frac{g \cdot \left(\frac{t}{2}\right)^2}{2} = \frac{g \cdot t^2}{8} \tag{3.30}$$

where h is jump height, t is full jump duration (both up and down).

Jumping observation revealed that in a continuous jumping scenario the time between hitting the ground and going up again takes approximately 40% of the actual jump time. This way full jump period is T = 1.4·t and jumping frequency is f = 1/T. The relationship between continuous jumping frequency and jumping height can be expressed as:

$$h = \frac{g \cdot \left(\frac{1}{1.4f}\right)^2}{8} = \frac{g}{15.68 \cdot f^2} \tag{3.31}$$

where h is jump height and f is continuous jumping frequency.

Taking into account the relationship given in Eq. (3.31) numerical analysis was performed. Because jumping frequency and jumping height is related it is purposeful to analyze only jump height impact towards the acceleration measurements as it is related to the jumping frequency and the relation is known.

Numerical analysis was performed on developed model to see what the impact of the jumping height and frequency is towards the measured accelerations. Analysis results are given in Table 3.14.

As results show relative differences for different jumping heights and frequencies are very similar. This implies the conclusion that jumping height and jumping frequency also are not main factors that influence acceleration measurement errors. It was showed before that belt tightening is not a factor also, meaning that only tissue rheological properties are the only important factor that influences the measurements. This conclusion has important practical value. It means that error reduction can be achieved only by having reduced human body surface tissue rheological model's properties.

**Table 3.14** Jumping height and frequency impact towards acceleration difference resulting from soft tissue impact

| Jump height, cm | Jumping frequency, Hz | Acceleration amplitude, m/s² (no soft tissue impact) | Modeled acceleration amplitude, m/s² (soft tissue impact is present) | Relative error, % |
|---|---|---|---|---|
| 5 | 3.6 | 8.079 | 9.035 | 11.83 |
| 5 | 3.6 | 15.970 | 17.830 | 11.65 |
| 13 | 2.2 | 9.681 | 11.170 | 15.38 |
| 13 | 2.2 | 19.370 | 21.990 | 13.53 |
| 20 | 1.8 | 8.065 | 9.040 | 12.09 |
| 20 | 1.8 | 17.440 | 19.410 | 11.30 |

# References

1. Ananthasures GK (2003) Optimal synthesis methods for MEMS. Kliuver, p 150
2. Mukherjee T, Zhou Y, Fedder GK (1999) Automated optimal synthesis of accelerometer. In: 12 IEEE International conference micro electro mechanical systems MEMS'99, Orlando, USA, pp 326–331
3. Yuan W, Chang H, Li W, Ma B (2006) Application of an optimization methodology for multidisciplinary system design of micro gyroscopes. J Microsyst Technol 12(4):315–323
4. Tay FE, Jun X, Logeeswaran VJ (2000) Optimization methodology for low-g micro accelerometer. J Micromach Microfabr 128:128–139
5. Pedersen CBW, SeshiaAA (2004) On the optimization of compliant force amplifier mechanisms for surface micromachined resonant accelerometers. J Micromech Microeng 14(10):1281–1293
6. Ostasevicius V, Gaidys R, Dauksevicius R (2009) Numerical analysis of dynamic effects of a nonlinear vibro-impact process for enhancing the reliability of contact-type MEMS devices. Sensors (Basel) 9(12):10201–10216
7. Tarabini M, Saggin B, Scaccabarozzi D, Moschioni G (2012) The potential of micro-electro-mechanical accelerometers in human vibration measurements. J Sound Vib 331(2):487–499
8. Lee H, Park JW, Helal A (2009) Estimation of indoor physical activity level based on footstep vibration signal measured by MEMS accelerometer in smart home environments. In: MELT'09 Proceedings of the 2nd international conference on mobile entity localization and tracking in GPS-less environments, pp 148–162
9. Bliley KE, Schwab DJ, Holmes DR et al (2006) Design of a compact system using a MEMS accelerometer to measure body posture and ambulation. In: Proceedings-IEEE symposium on computer-based medical systems, pp 335–340
10. Lyons RG (2011) The discrete Fourier transform. Windowing. In: Understanding Digital Signal Processing, pp 89–95
11. Lyons RG (2011) Finite impulse response filters. An introduction to finite impulse response (FIR) filters. In: Understanding Digital Signal Processing, pp 170–175
12. Lyons RG (2011) Infinite impulse response filters. A brief comparison of IIR and FIR filters. In: Understanding Digital Signal Processing, pp 332–333
13. Armin G (1997) Fundamentals of videogrammetry—a review. Hum Mov Sci 16:155–187
14. Bouten CVC, Koekkoek KTM, Verduin M, Kodde R, Janssen JD (1997) A triaxial accelerometer and portable data processing unit for the assessment of daily physical activity. IEEE Trans Biomed Eng 44:136–147
15. Benevicius V, Ostasevicius V, Gaidys R (2013) Human body rheology impact on measurements in accelerometer applications. J Mech/Mech 19(1):40–45
16. Ledoux WR, Hillstrom HJ (2001) Acceleration of the calcaneus at heel strike in neutrally aligned and pes planus feet. Clin Biomech (Bristol, Avon) 16(7):608–13
17. Qu H, Fang D, Xie HA (2008) Monolithic CMOS-MEMS 3 axis accelerometer with low noise, low power dual chopper amplifier. IEEE Sensors J 8(9):1511–1518
18. Reilly SP, Leach RK, Cuenat A, Awan SA, Lowe M (2006) Overview of MEMS sensors and the metrology requirements for their manufacture, NPL Report DEPC-EM 008
19. Yoshida K, Matsumoto Y, Ishida M, Okada K (1998) High-sensitive three axis SOI capacitive accelerometer using dicing method. In: 16th sensor symposium, pp 25–28
20. Benevicius V, Ostasevicius V, Venslauskas M, Dauksevicius R, Gaidys R (2011) Finite element model of MEMS accelerometer for accurate prediction of dynamic characteristics in biomechanical applications. J Vibroeng/Vibromech (Lithuanian Academy of Sciences, Kaunas University of Technology, Vilnius Gediminas Technical University. Vibromechanika, Vilnius) 13(4):803–809

21. Benevicius V, Ostasevicius V, Gaidys R (2013) Identification of capacitive MEMS accelerometer structure parameters for human body dynamics measurements. Sensors 13 (9):11184–11195
22. Müller-Riemenschneider F, Reinhold T, Berghöfer A, Willich SN (2008) Health-economic burden of obesity in Europe. Eur J Epidemiol 23(8):499–509
23. WHO/Europe (2011) Home. http://www.euro.who.int/en/home [2011/06/13]
24. Neck support collar. http://www.soospine.com/images/1miami_jB.jpg
25. Benevicius V, Gaidys R, Ostasevicius V, Marozas V (2014) Identification of rheological properties of human body surface tissue. J Biomech 47(6):1368–1372
26. Markidou A, Shih WY, Shigh W (2005) Soft-materials elastic and shear moduli measurement using piezoelectric cantilevers. Rev Sci Instrum 76(6):7
27. Gennisson JL, Baldeweck T, Tanter M, Catheline S, Fink Mathias, Sandrin L, Cornillon C, Querleux B (2004) Assesment of elastic parameters of human skin using dynamic elastography. IEEE Trans Ultrason Ferroelectr Freq Control 51(8):980–989
28. Mukherjee S, Chawla A, Mohan D, Metri M (2006) Modeling of body parts consisting of bones as well as soft tissue: an experimental and finite element study. In: IRCOBI conference, Lisbon, Portugal, pp 367–368

# Chapter 4
# MOEMS-Assisted Radial Pulse Measurement System Development

**Abstract** Blood pressure is the first assessment of blood flow and still is the easiest parameter to measure. Examination of wrist radial pulse is non-invasive diagnostic method, which takes a very important place in traditional Chinese medicine. Hence, it lacks the consistency and reliability, which limits practical application in clinical medicine. Thus, the quantitative characteristics of the method of the wrist pulse diagnosis is a prerequisite for its further development and wide use. Noninvasive CCD sensor based on a hybrid measurement systems for the analysis of the radial pulse signals is developed. Various input pressure and fluids of different viscosity are used in the assembled artificial circulation system to test the effectiveness of laser triangulation technology to improve the detection sensitivity by a microfabricated MOEMS placed on a synthetic vascular graft and on the human wrist. MOEMS design methodology is presented. The simulation results provide the right to choose the form of the membrane. Technological process of formation of MOEMS is performed using advanced microfabrication technology. In "in vitro" and "in vivo" measurement results are adequate and suggest practical relevance of the proposed methodology.

## 4.1 Validation of Noninvasive MOEMS-Assisted Radial Pulse Analysis System

### 4.1.1 Importance of Blood Pressure and Radial Pulse Diagnosis

Measurement of blood pressure is one of the most important measurements of health [1]. Blood pressure is the first assessment of blood flow and still is the easiest parameter to measure. Thus, it is a convenient measure of the patient's health, as well as one of the most important vital functions. The importance of measuring blood pressure is that it is firmly connected, via the variable impedance of human

© Springer International Publishing AG 2017
V. Ostasevicius et al., *Biomechanical Microsystems*, Lecture Notes
in Computational Vision and Biomechanics 24,
DOI 10.1007/978-3-319-54849-4_4

organs, with the physiology of the human body, and nearly all physiological processes that are carried in blood pressure signals, either arterial or venous. Additionally, some features of arterial blood pressure, such as mean arterial pressure, systolic and diastolic pressure, are epidemiologically associated with numerous circulatory system diseases such as hypertension, obesity, epilepsy and cancer, myocardial infarction, stroke, congestive heart valve failure, atherosclerosis, and as it regards the problems associated with kidney disease and diabetes [2].

The significance of blood pressure measurements can be soundly stressed by epidemiological data. In the whole world extremely big number of people suffers from arterial hypertension, just in USA 56 million of patients have hypertension problems [1]. The biggest advantage in the field of MOEMS MEMS-devices open up entirely new possibilities for more accurate, permanent and blood pressure in real-time measurements.

### 4.1.2   Metrology of Arterial Blood Pressure

The arterial blood pressure signal originates in the left ventricle and propagates through the systemic arteries, changing its static and dynamic parameters. The position of the heart is considered to be the reference point in which the static component (hydrostatic pressure) is assumed to be zero. Blood pressure parameters vary from subject to subject and undergo changes according to the physiological state as well as a wide range of diseases. Nevertheless, some average numbers measured on the healthy population can be given. The arterial blood pressure amplitude for a healthy person, measured in the aorta, varies from 40 to 80 mmHg; which corresponds to 120–140 mmHg at the maximum of the pulse waveform, the so-called systolic pressure, and to 80–90 mmHg at the minimum, the so-called diastolic pressure. The frequency of the signal varies from 40 to 200 beats per minute [2]. Thus, the period is in the range from 1.5 to 0.3 s. The average period for a resting healthy subject is about 0.85 s. The generic shape of the typical signal, recorded in a main artery, is depicted in Fig. 4.1. The first rising portion of the pulse has an average dP(t)/dt around 200 mmHg/s, and the falling part dP(t)/dt is around 130 mmHg/s, where P(t) is the blood pressure in time domain. The typical range of amplitudes does not cover the pressure values that can be generated by a hydrostatic pressure. Body position can significantly shift the blood pressure in the aorta arch by as much as 40 mmHg in a standing position, or even 80 mmHg in a reversed position, with head down [1]. Also, body movements or exercise such as running can add to the amplitude by 60 mmHg [2]. Aging also increases the average systemic pressure, shifting it by about 1 mmHg per year and slightly increasing the peak-to-peak amplitude of the pulse.

Many authors treat the blood pressure signal as a periodic one and analyze it by using the Fourier transform:

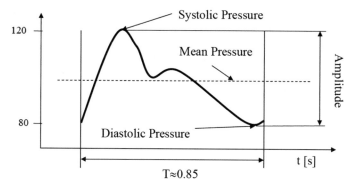

**Fig. 4.1** Generic arterial blood pressure cycle in time domain [1]

$$F(\omega) = \frac{1}{2\pi} \int\limits_{-\infty}^{\infty} BP(t) \exp(-j\omega t) dt, \qquad (4.1)$$

where $F(\omega)$ is a Fourier transform, $BP(t)$ is blood pressure function in time domain, j is an imaginary number, and $\omega$ is angular frequency. This approach offers a very convenient set of tools allowing analysis of the signal for specific frequencies— harmonics, by using only two numbers: amplitude of the harmonic and its phase. If the system is assumed to be linear and periodic, the Fourier analysis can provide the complete description of the system dynamics. The fundamental period is often assumed to be the heart cycle or the multiplication of it. Many researches use the Fourier methodology, such as spectrum analysis, to express the blood signal pressure features [1, 2]. Typical blood pressure spectral content is shown in Fig. 4.2. In fact, blood pressure is not a pure periodic signal. The long-term extensive studies on Heart Rate Variability (HRV) showed the short- and long-term fluctuations in the HR fundamental frequency. Also, the spectral content of the

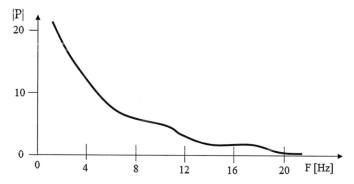

**Fig. 4.2** Arterial blood pressure in frequency domain (Spectrum)

blood pressure signal undergoes changes due to physiological responses of the cardiovascular system. In addition, many types of artifacts are involved, introducing stochastic or rather chaotic components. The situation is made worse with the technical limitations to the digital signal analysis such as a finite period of collecting data, and a finite period of sampling (sampling frequency) data or quantization effect. In a real experiment, the "ideal" continuous Eq. 4.2 has to be supplemented by a discrete one [1]:

$$F(k) = \frac{1}{N} \sum_{n}^{N-1} BP(n) \exp\left(-j\frac{2\pi nk}{N}\right), \tag{4.2}$$

where F(k) is a Fourier transform, BP(n) is blood pressure function in a discrete time domain, j is an imaginary number, k is an angular frequency, n is the number of sample, and N is the total number of samples.

Thus, the conversion from time domain to frequency domain utilizing the formula will introduce an error due to sampling and an energy leakage. The energy leakage is related to the finite size of the analyzing window—$N\Delta t$ where N is number of samples and $1/\Delta t$ is sampling frequency [1]. Because the analyzing time window has a finite size, by definition it cannot cover all low frequencies. Also, the analyzing window can generate some high frequency artifacts by unmatched signal levels at the ends of the window. The sampling rate limits the frequency range to be analyzed unless the Nyquist criterion is met: $1/\Delta t > 2$ fmax. Where $1/\Delta t$ is sampling frequency and fmax is the maximum frequency present in the signal spectrum.

In the arterial measurements pressure pulse contains from 6 to 20 harmonics [1].

### 4.1.3   Radial Pulse Diagnosis

Chinese pulse diagnosis in traditional Chinese medicine (TCM) has been practiced for over 2000 years [3]. Without any technological equipment of Chinese medicine employed the fingertips to feel the wrist pulses patients to determine their health status. Depending on the hand and wrist pulse sensing, doctor can determine the status of the various organs of the patient (Fig. 4.3).

Wrist pulse was seen as the most fundamental of life signals, containing basic information about the health of the individual. In traditional Chinese pulse diagnosis theory (TCPD) pulse signals, which is caused by fluctuations in the flow of the radial artery, contain a rich and important information that can reflect the state of human organs, i.e., gallbladder, kidney, stomach, lung, and so on. This pathological change of internal organs may be reflected by variations in the rhythm, the speed, the radial pulse power, by which experienced practitioners can tell the state of human health [3]. Furthermore, TCPD is non-invasive and convenient for efficient diagnosis. Clinical studies demonstrate that in patients with hypertension,

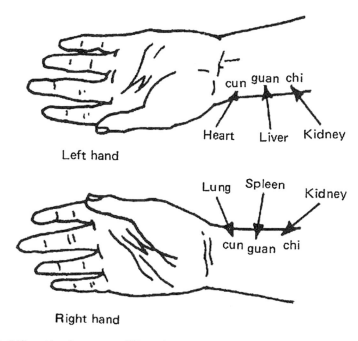

**Fig. 4.3** Different hand represents different human organ

hypercholesterolemia, cardiovascular disease and diabetes exhibit premature loss of elasticity of arteries and endothelial dysfunction, which eventually led to a decrease in flexibility and an increased load on the vascular circulatory system. The shape of the wrist pulse, amplitude and rhythm also changed according to the hemodynamic characteristics of blood flow [3–8].

### 4.1.4   Radial Pulse Characteristics

The beating heart creates pressure and flow of waves, which spread throughout the arterial system. The shapes of wrist pulse signals change by their continuous interaction with irregular arterial system. The pressure waves are expanding the arterial wall while traveling, and decompositions are recognized as the wrist pulse. Each gap reflects incident waves into mechanical and geometrical properties of the arterial tree, such as the bifurcation and stenosis. Sensible wrist pulses can thus be understood in terms of a forward traveling wave component, collective waves coming from the heart to the periphery and containing information of the heart; and one backward traveling wave component, the collective waves containing information about the places of reflection, i.e. kidneys, stomach, spleen, liver, lungs, etc. In addition, the reflected pressure waves tend to increase the load on the heart and

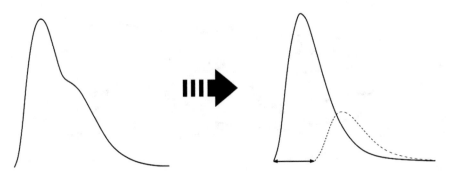

**Fig. 4.4** Forward wave is higher in amplitude, and backward wave is lower in amplitude, with a phase shift [4]

play a major role in determining the pattern of wrist pulse waveforms. Consequently, wrist pulse signals can be expressed in terms of its forward and backward components running with a phase shift in time as shown in Fig. 4.4 [4].

Normal wrist pulse waveform has a sharp enough, to move up to the moment of peak steady, high speed down stroke and decay. The reflected wave is also similar to the original wave shape, but smaller in amplitude. Young healthy people tend to have a pulse patterns as shown in Fig. 4.5 [5]. Following a traditional Chinese

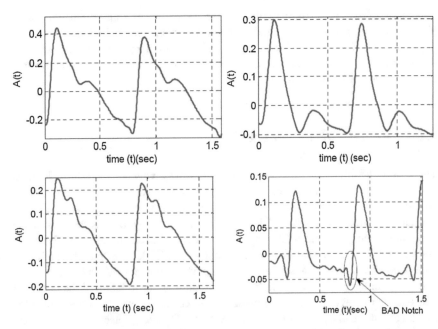

**Fig. 4.5** Radial pulse patterns: *top-left* taut pulse pattern; *top-right* slippery pulse pattern; *bottom-left* moderate pulse pattern; *bottom-right* abnormal pulse pattern with BAD Notch [5]

medicine terminology graphs clearly show the presence of the dicrotic notch and dicrotic wave and impulses that can be identified as a tight, slippery or moderate.

The abnormal pulse pattern shows formation of a unique 'V' shaped notch identified as BAD Notch (Fig. 4.5 bottom-right).

Such non-invasive, biocompatible and portable MOEMS device could easily find the place in up to date market and would be extremely helpful in ambulance medicine, not to mention athletes, older persons and patients which need to have real-time health monitoring.

## 4.2 Micro Membrane Design

### 4.2.1 Evaluation of Residual Stresses

All micro-machined elements tend to have residual stresses that can be vital for the operation of the micro sensor. Especially, if the signal is registered using optical sensor, like in the presented case, there is a need to examine the surface of micro-element in details. Even small part of membrane bow can lead to the distortion of signal registration, leaving micro sensor totally inoperable. It is often observed that stresses develop in films during deposition or growth at elevated temperatures. These stresses arise, because generally films are deposited under non-equilibrium conditions.

#### 4.2.1.1 Stresses in Thin Polysilicon Film Formed During Vapor Deposition

For thin coating, misfit strain is usually assumed to be dealt with fully in it, so that the stress level is easily obtained by multiplying by two-axis module [9]. For all types of coating, the main sources of residual stresses are differential thermal contraction and the phenomena occurring during the deposition. It is common for them to be called respectively as external and internal stresses. Other processes such as phase transformations, plastic flow, creep, etc., can effectively generate a strain mismatch. They can, in principle, be handled in basically the same manner as the thermal stress. In the simulation, significant attention has been paid to differential thermal contraction stresses.

#### 4.2.1.2 Differential Thermal Contraction Stresses

Thermal stresses can readily be calculated from knowledge of the thermal expansivities of the constituent materials. The associated misfit strain can be written as [10]:

$$\Delta\varepsilon = \int_{T_1}^{T_0} (\alpha_s - \alpha_f)dT, \qquad (4.3)$$

where $\Delta T$ is the temperature change between temperature in which deposition process took place and ambient temperature, $\alpha_s$-coefficient of thermal expansion of substrate, $\alpha_f$-coefficient of thermal expansion of thin film. This strain, and the associated stress level in the coating at room temperature will obviously be more when deposition occurs at high temperatures and when large expansivity mismatch. While this equation is commonly used, it should be noted that the neglect of the temperature dependence of expansivities may be inaccurate. Misfit strain at an ambient temperature $T_0$ after cooling from deposition temperature $T_1$ to be obtained from:

$$\Delta\varepsilon = \int_{T_1}^{T_0} (\alpha_s - \alpha_f)dT. \qquad (4.4)$$

### 4.2.1.3  Deposition (Intrinsic) Stresses

This type of stress can occur in several ways, during the molecular vapor deposition. For example, molecular species entering with high energies can become implanted within the field, where they may occupy interstices (or free crystal lattice sites that are thermodynamically stable) and thus generate a compressive load. The bombardment of energetic, non-depositing species may also contribute to such a site of employment and therefore have the same effect, which is often called "atomic peening". On the other hand, the processes can also take place during the deposition, which generates excess vacancies and hence tensile stress. For example, preferential removal of deposited molecular species by etching immediately after the deposition may lead to such an effect. Another point should be noted that there is a tendency to excess vacancies and interstitials/vacancies depletion to be removed by short-range (surface) diffusion in the presence of a sufficient amount of heat energy available. Both substrate temperature and a short duration of thermal energy injected or implanted by species bombardment are important in determining the extent to which annealing processes occur.

In all cases, stresses arise mainly as a result the material is deposited at first in a non-equilibrium state and the differential thermal contraction occurring between the coating and the substrate during temperature changes after deposition. Both mechanisms may generate either tensile or compressive stress in the coating. While the differential thermal expansion stress is fairly easy to predict the mechanisms that determine the deposition voltage may be more complex.

This is the reason why during modeling step thermal expansion stresses are chosen, since the problem could be described by the tools of finite element

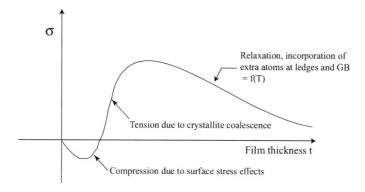

**Fig. 4.6** Average film stress as a function of film thickness [10]

(FE) modeling in not significantly difficult way. In order to predict and explain the fabrication and modeling results typical behavior of the average film stress as a function of film thickness shown in Fig. 4.6 was chosen as a reference.

As this work started with the fabrication of micro-membranes as basic sensing elements of MOEMS, it was essential to examine why fabrication process was not successful in some cases and employing capabilities of finite element modeling programs to find out what would be the best parameters and best geometric shape of micro-elements in order to avoid or reduce unwanted phenomenon.

### 4.2.2 Three Dimensional Finite Element Model of Micro-membrane

Micro-membrane, micro-membrane matrixes or micro mirror arrays are generally found as the main structural elements in the sensors, MOEMS, actuators, filters, etc. Structural element usually consists of silicon dioxide, silicon nitride, polycrystalline silicon thin film deposited on a thick silicon wafer. In particular, the final process of manufacturing a structural element should be uniform and even composed only of silicon dioxide/polysilicon connection. But because of the possible not uniform etching process, some impurities or silicon dust may be present in the structure. However, in the simulation it is approximated that the micro-membrane consists of pure silica/polysilicon connection.

#### 4.2.2.1 Modelling Steps

The goal of finite element analysis is to recreate the mathematical behavior of the real technical system. This means that the mathematical model of a physical prototype will be generated. Currently, there are various software Fe (Adina, ANSYS,

COMSOL Multiphysics, Nastran, SIMULIA, etc.), which are capable of division of the object into a plurality of elements and components, assigning materials, change their properties, the use of the boundary conditions. All these features help to create a realistic model and simulate the behavior under the influence of ambient conditions. Nevertheless, it must be emphasized that the general and special FE modeling includes the following stages:

(1)  Geometry development;
(2)  Material and its properties assignment;
(3)  Mesh generation;
(4)  Imposition of boundary and loading conditions;
(5)  Run analysis;
(6)  Plotting the static or dynamic solution.

Geometric modeling silica/polysilicon micro-membranes was performed using COMSOL Multiphysics. Various geometric patterns were obtained for different sizes and shapes of the structures. The square structure of the membrane after the micro-fabrication process, and the one introduced for the simulation are shown in Fig. 4.7.

As significant interest has been received by micro membrane and its behavior due to thermal stresses, thus simplifying the square object diagram with typical parameters used for numerical simulation are shown in Fig. 4.8 (the same material properties for a circular membrane, a). Mechanical model of square and round micro-membrane has been created using the same finite element (FE) simulation software COMSOL Multiphysics [11]. Here, the IP model describes the dynamics of the microstructure and follows the classical equation of motion represented in the general form of a matrix:

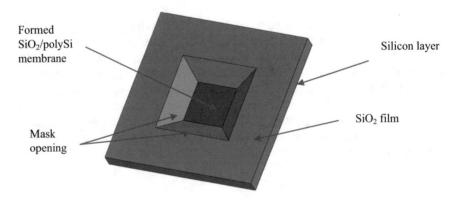

**Fig. 4.7**  Geometric model of SiO$_2$/polySi membrane (*darksquare* at the centre)

$$[M]\{\ddot{U}\} + [C]\{\dot{U}\} + [K]\{U\} = \{Q(t, U, \dot{U})\}, \tag{4.5}$$

where $[M], [C], [K]$ the mass, damping and stiffness matrices, respectively; $\{\ddot{U}\}$, $\{\dot{U}\}$, $\{U\}$ displacement, acceleration and velocity vectors, respectively; $\{Q(t, U, \dot{U})\}$-vector representing the sum of the forces acting on the micro-membrane.

Table 4.1 defines exact parameters and boundary conditions.

Analyzing figure presented above should be understood that the simulated micro-membrane was fixed around the perimeter just leaving free translational movement in the direction Z., i.e. free translational movement was possible in only one direction.

It should be understood that this part of numerical modeling is based on fabrication peculiarities of micro objects. The application for biomedical application of such micro membranes as basic parts of optical sensor will be discussed in computational fluid dynamics section.

## 4.2.3   Square Membrane Modeling

Usually, square elements are preferable for the pressure sensor geometry, because the high stress is generated by the applied pressure load resulting in high sensitivity. Furthermore, in the particular case of a square geometry it is desirable for a better and easier target light and reflection. Nevertheless, as this chapter discusses the residual stresses the essential question to be answered: is the shape of micro-fabricated object essential to the residual stress point of view? Should no only a square membrane be modeled and analyzed? Using numerical simulation software COMSOL first step is to find the natural frequencies of the known

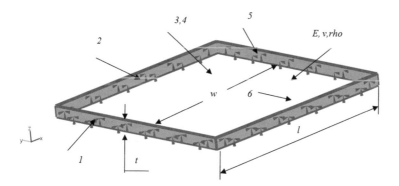

**Fig. 4.8**   Schematic representation of a micro-membrane

**Table 4.1** Physical, mechanical properties and boundary conditions of micro-membrane

| Description and symbol | Value | Unit |
|---|---|---|
| Length l | 0.4, 1, 5 | mm |
| Width w | 0.4, 1, 5 | mm |
| Thickness t | 20 | μm |
| Young's modulus E | 155, 130 | GPa |
| Density ρ | 2330, 2200 | kg/m$^3$ |
| Poisson's ratio $v$ | 0.16, 0.23 | – |
| Room temperature $T_0$ | 20 | °C |
| Deposition temperature $T_1$ | 600 | °C |
| Residual stress $\sigma_r$ | 50 | MPa |
| Residual strain $\varepsilon$ | $\sigma_r(1 - v)/E$ | – |
| Coefficient of thermal expansion (1/K) | $\varepsilon/(T1 - T0)$ | – |
| Boundary conditions | 1,2,5,6—fixed planes 3,4—free planes | – |

**Table 4.2** Simulated natural frequencies of square membranes

| Side length, (mm), 20μm thickness | Natural frequency values (kHz), 6 eigenmodes | | | | | |
|---|---|---|---|---|---|---|
| 0.4 | 1080 | 2058 | 2058 | 3896 | 5689 | 6175 |
| 1 | 177 | 365 | 384 | 758 | 856 | 1034 |
| 5 | 7.3 | 23.9 | 23.9 | 36.4 | 43.3 | 43.5 |

geometry of square membranes. Are the natural frequencies or the eigenfrequency are frequencies at which the oscillating system can vibrate?

Here, the side lengths of membranes were chosen to be 0.4, 1 and 5 mm. Thickness of them 20 μm. Table 4.2 represents all the values of eigenfrequencies resulting from simulation.

Figure 4.9 represents the shapes of eigenmodes of 5 mm side length membrane for interest. The remaining dimension membranes has very similar characteristics of eigenmodes, thus it was no need to present it in details.

Further, modeling continued taking into account the phenomenon of differential thermal stresses. Here, when the assembly is cooled to room temperature, the substrate film shrank differently and caused strain in the film. Thus, this analysis shows how the thermal residual stress changes the resonance frequency of the structure. Assuming that the material is isotropic, the stress is constant throughout the film thickness and the stress component in a direction normal to the substrate is zero. Then, relationship between stress-strain is:

$$\sigma_r(1 - v)/E, \tag{4.6}$$

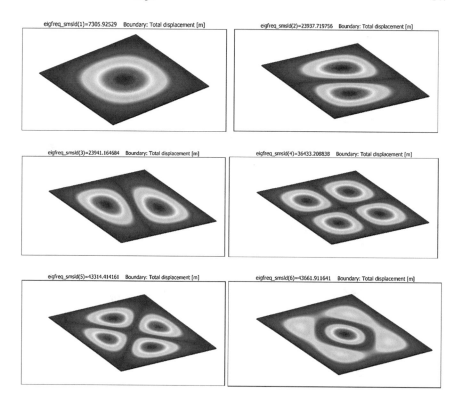

**Fig. 4.9** Eigenmodes of 5 mm side length membrane from: *top-left* to *bottom-right*—six eigenmodes respectively

where in the Young's modulus $E$, Poisson's ratio $v$, strain $\varepsilon$ is given by:

$$\varepsilon = \Delta\alpha\Delta T, \tag{4.7}$$

where $\Delta\alpha$ the difference between the coefficients of thermal expansion, and $\Delta T$ the difference between the deposition temperature and the normal operating temperature.

To evaluate the residual thermal stresses in the temperature differences in particular case were 600 °C (usual deposition temperature present during polysilicon vapor deposition) and ambient room temperature of 20 °C. Static analysis and von-misses stress distribution is presented as well.

Tables 4.3, 4.4 and 4.5 represent the three different dimension square membrane simulation. Comparison of natural frequencies of the thermally stressed structure and disregarding them is presented.

**Table 4.3**  0.16 mm$^2$ area membrane

| Natural frequencies without thermal stresses | Natural frequencies with thermal stresses |
| --- | --- |

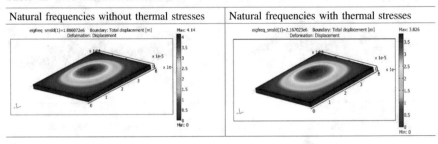

**Table 4.4**  1 mm$^2$ area membrane

| Natural frequencies without thermal stresses | Natural frequencies with thermal stresses |
| --- | --- |

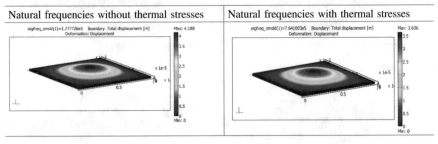

**Table 4.5**  25 mm$^2$ area membrane

| Natural frequencies without thermal stress | Natural frequencies with thermal stress |
| --- | --- |
| Eigenfrequency is 7.3 kHz | Eigenfrequency is 146 kHz |
| The deformed shape of the geometry of the membrane under thermal stress | Von misses stress distribution passing through the center of the membrane |

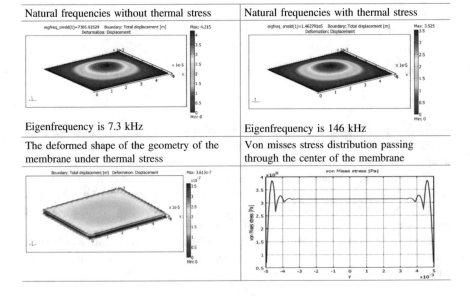

### 4.2.4 Circular Membrane Modeling

Circular membrane has the lowest load at its edges when applying the same pressure than a square diaphragm, but the largest center of deflection can be seen in the circular diaphragm. Thus, in applications, the maximum deviation is paramount to the circular diaphragm offered. Circular diaphragm is most favored from the design engineering point of view.

As it was mentioned above the essence of this section is to discover if shape of micro object has some effect on thermal stresses, therefore modeling of circular membranes took place.

Employing mathematical formulation of circle area $A = \pi R^2$, radius of membranes was found for further numerical simulation and comparison.

In order to have the same area of a circular object as square ones, there was a need to solve simple equation.

$$R^2 = \frac{A}{\pi}, \tag{4.8}$$

where R is radius of membrane, A is area of square membrane, $\pi$ is mathematical constant. Substituting values of the area of square membranes radius of circular membranes were found to be 0.225, 0.564, and 2.822 mm from smallest to biggest respectively. Table 4.6 represents the six values of eigenfrequencies resulting from simulation.

Tables 4.7, 4.8 and 4.9 represent the three different area circular membrane simulations. Comparison of natural frequencies of the structure with thermal stresses and disregarding them is presented. Moreover, static analysis of the deformed shape of the membrane under the thermal influence and von-misses stress distribution is delivered.

After employing numerical simulation of micro membranes of different shapes under the influence of thermal stresses and disregarding them, it became obvious that the shape shouldn't be the concern for this type of analysis. Tables 4.10 and 4.11 represent resonant frequencies with and without residual stress of square and circular membranes respectively.

Based on simulation results, it can be seen that with a smaller diaphragm and the same thickness the impact of residual stresses on the membrane decreases as the membrane area is reduced. Comparing the resonant frequency of small membranes,

**Table 4.6** Simulated natural frequencies of circular membranes

| Radius (mm), 20μm thickness | Natural frequency values (kHz), 6 eigenmodes | | | | | |
|---|---|---|---|---|---|---|
| 0.225 | 498 | 958 | 959 | 1401 | 5157 | 5183 |
| 0.564 | 80.3 | 180.2 | 180.2 | 356 | 458 | 589 |
| 2.822 | 3.26 | 8.8 | 8.8 | 15.5 | 17.6 | 20.5 |

**Table 4.7**  0.16 mm$^2$ area membrane

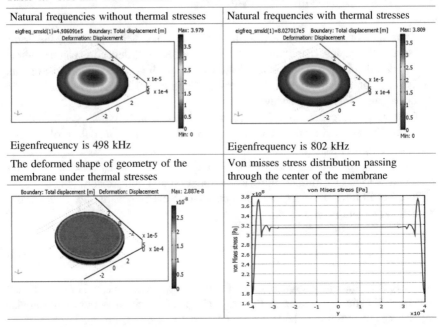

| Natural frequencies without thermal stresses | Natural frequencies with thermal stresses |
|---|---|
| Eigenfrequency is 498 kHz | Eigenfrequency is 802 kHz |
| The deformed shape of geometry of the membrane under thermal stresses | Von misses stress distribution passing through the center of the membrane |

**Table 4.8**  1 mm$^2$ area membrane

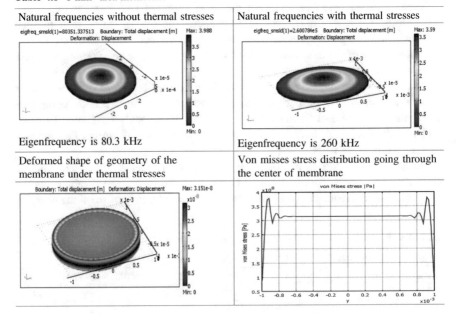

| Natural frequencies without thermal stresses | Natural frequencies with thermal stresses |
|---|---|
| Eigenfrequency is 80.3 kHz | Eigenfrequency is 260 kHz |
| Deformed shape of geometry of the membrane under thermal stresses | Von misses stress distribution going through the center of membrane |

**Table 4.9**  25 mm$^2$ area membrane

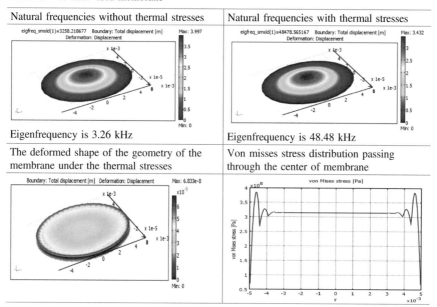

| Natural frequencies without thermal stresses | Natural frequencies with thermal stresses |
|---|---|
| Eigenfrequency is 3.26 kHz | Eigenfrequency is 48.48 kHz |
| The deformed shape of the geometry of the membrane under the thermal stresses | Von misses stress distribution passing through the center of membrane |

**Table 4.10**  Resonant frequencies with and without residual stresses of square micro-membrane of different side lengths

| Area of membrane | 0.16 mm$^2$ | 1 mm$^2$ | 25 mm$^2$ |
|---|---|---|---|
| Resonant frequency without stress (kHz) | 1080 | 177 | 7.3 |
| Resonant frequency with stress (kHz) | 2160 | 764 | 146 |

**Table 4.11**  Resonant frequencies with and without residual stresses of circular micro-membrane of different areas

| Area of membrane | 0.16 mm$^2$ | 1 mm$^2$ | 25 mm$^2$ |
|---|---|---|---|
| Resonant frequency without stress (kHz) | 498 | 80.35 | 3.26 |
| Resonant frequency with stress (kHz) | 802 | 260 | 48.48 |

it can be seen that the solution of problems, including residual stresses resonant frequencies differ less than twofold in both cases. Thus, the thermal load on the square millimeter membrane even more than 3–5 times make a difference to eigenmodes of structure for circular and square membrane, respectively. The resonant frequency of 25 mm membrane area, including the thermal load is already given rise even 14–20 times. Von misses stress distribution is most noticeable near the point of attachment of micro-devices. Thus it is clear that in order to properly micro-fabricate operable membrane area ratio, width and the micro-device should be as small as possible. Such ratio will also serve as an advantage for better sensitivity of the micro element.

## 4.3   Micro Membrane Fabrication and Experimentation

MOEMS are commonly fabricated from silicon and its compounds, silicon (di) oxide, and silicon nitride [12–14]. MOEMS typically require optical surfaces with high flatness and low roughness to be combined with high quality mechanical parts and low power [15].

In the specific case flatness and mirror finish surface are most important aspects for device operation, since the displacement of the point of interest will be registered using laser triangulation displacement sensor. Therefore, polysilicon layer was used as reflective coating and micro membrane.

From the application point of view following tasks for manufacturing micro-objects were formulated:

Fabricate micro-membranes with mirror finish surface.
Fabricate micro-membranes of different dimensions.
Fabricate micro-membranes of different geometries.
Fabricate micro-membranes with smallest possible residual stresses.

Following tasks step by step it was obvious that in order to obtain desired result quite complex process should take place. Primary assumption of process was created that freely hanging polysilicon micro membranes with optical grating will be obtained by surface and bulk micromachining technologies. Procedure of formation of micro membrane and optical grating is presented in Fig. 4.10.

For the $Si_3N_4$ deposition layer surface micromachining technology was used. To form the optical grating, bulk micromachining technology used. During the etching process, the upper side of the plate is covered with low stress transparent $Si_3N_4$, where using the RIE (reactive ion etching) of the diffraction grating techniques must be formed (also transparent to infrared radiation). The principle is simple membrane formation. With a silicon substrate/silicon dioxide thickness of 300 μm, a thin film (20 μm) is deposited on the polysilicon wafer by pyrolysis (thermal decomposition) silane, $SiH_4$, a low pressure inside the reactor 25–130 Pa at a temperature of 580–650 °C. This pyrolysis process includes the following basic reaction: $SiH_4 \rightarrow Si + 2H_2$. Polysilicon deposition rate increases rapidly with increasing temperature, since it follows the Arrhenius equation:

$$R_{ar} = Ae^{-qE_a/kT}, \tag{4.9}$$

where $R_{ar}$ the deposition rate, $E_a$ is the activation energy in electron volts, $T$ is the absolute temperature in degrees Kelvin, $k$ is Boltzmann constant, $q$ is the electron charge, and $A$ is a constant. The activation energy for the deposition of polycrystalline silicon is about 1.7 eV.

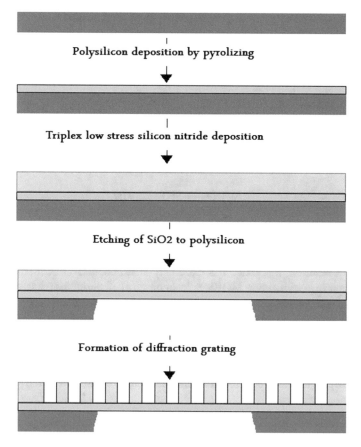

**Fig. 4.10** Schematics of process for the formation of a micro-membrane

## 4.3.1   Determination of Primary Data for Analyzed Objects

While starting the determination of primary data it was quite tricky to find out optimal geometries for micro-membrane, since the overall production process should become cost effective with minimum possible discrepancies. Nevertheless, decision was made to fabricate micro membranes of different geometrical parameters and different shapes (square and circular). Shortly, the dimensions of micro-membranes were chosen: square membranes with side lengths 0.4, 1 and 5 mm and circular membranes with radiuses 0.225, 0.564, and 2.822 mm respectively. For the consistency reasons thickness of the membranes there chosen to be constant in all the cases, namely 20 μm. It should be noted that finite element modeling of microstructure is practically impossible with used multiphysical modeling programs as COMSOL if area to thickness ratio is too big.

### 4.3.2  Deposition of Silicon Dioxide and Polysilicon

Silicon dioxide usually can be employed as structural or sacrificial layer in various microelectromechanical devices. In particular case it is used as sacrificial layer in order to obtain polysilicon membranes. $SiO_2$ can be obtained by the following methods: thermal oxidation, anodizing in liquid, chemical vapour deposition or plasma oxidation. Nevertheless, thermal oxidation is the most frequently applied [13]. In this case silicon oxidises easily when reacting with oxygen or water steam (at 1 atm) at elevated temperatures $(600 \div 1250)$ °C. Chemical reactions of outer Si wafer layer conversion into high quality $SiO_2$ can be written as [13]:

$$Si + O_2 \rightarrow SiO_2 \ (Dry);  \qquad (4.10)$$

$$Si + 2H_2O \rightarrow SiO_2 + 2H_2O \ (Wet).  \qquad (4.11)$$

$SiO_2$ has desired properties of thermal and electric insulation as well as chemical stability. In addition, this oxide has extremely small coefficient of thermal expansion (comparing with other MOEMS technological materials) and is capable of forming an almost perfect electrical interface with its substrate. Finally, analyzed compound is stable in water and at elevated temperatures giving opportunity to manipulate with it in terms of processing and application. Due to its familiarity, versatility and reliability $SiO_2$ is so widely-spread in microfabrication.

Main mechanical, material and thermal characteristics of silicon, silicon dioxide among other significant technological materials are presented in Table 4.12.

**Table 4.12** Main mechanical, material and thermal characteristics of significant technological materials [13]

| Material | Yield strength (GPa) | Young's modulus (GPa) | Density $(10^3 \ kg/m^3)$ | Thermal conductivity at 300 K (W/cm K) | Thermal expansion $(10^{-6}/°C)$ |
|---|---|---|---|---|---|
| Diamond | 53 | 10.35 | 3.5 | 20 | 1.0 |
| Si (SC) | $2.6 \div 6.8$ | 190 $\langle 111 \rangle$ | 2.32 | 1.56 | 2.616 |
| GaAs (SC) | 2.0 | 0.75 | 5.3 | 0.81 | 6.0 |
| $Si_3N_4$ | 14 | 323 | 3.1 | 0.19 | 2.8 |
| $SiO_2$ (fibers) | 8.4 | 73 | 2.5 | 0.014 | $0.4 \div 0.55$ |
| SiC | 21 | 448 | 3.2 | 5.0 | 4.2 |
| Al | 0.17 | 70.0 | 2.7 | 2.36 | 25.0 |
| AlN | 16.0 | 340 | 3.26 | 1.6 | 4.0 |
| $Al_2O_3$ | 15.4 | 275 | 4.0 | 0.5 | $5.4 \div 8.7$ |
| Stain. steel | $0.5 \div 1.5$ | $206 \div 235$ | $7.9 \div 8.2$ | 0.329 | 17.3 |

*SC* Single crystal, $\langle 111 \rangle$—crystallographic plane orientation

**Polysilicon layer 20μm**

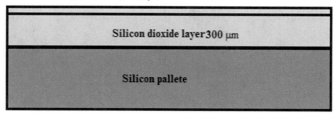

**Fig. 4.11**  Deposition of silicon dioxide and thin film polysilicon layers

In our particular case fabrication sequence starts with deposition of silicon dioxide layer (300 μm thick) on silicon palette to improve adhesion and afterwards using surface micromachining polysilicon thin film (20 μm thick) is deposited on the silicon dioxide sacrificial layer (Fig. 4.11). In contrast to bulk micromachining, in which microstructures are formed by etching a bulk substrate surface micromachining creates patterns by adding materials, layer by layer, on the substrate surface. The thin film layers are deposited, typically 15–25 μm thick, serving as some of the structural layer and the other layer being removed. Dry etching is used to define layers of structure and the final wet etching step frees them from the supporting substrate by removing the sacrificial layer.

Before any deposition process the specimens should be properly prepared in order to reduce the amount of dirt on it (dust, oxides). In particular case silicon palettes was cleaned immersing wafers into boiling acetone and keeping it inside acetone for 120 s (see Fig. 4.12).

Furthermore, in order to make the surface of palettes as clean as possible all specimens are to be cleaned using the help of plasma cleaning techniques.

Typically, plasma treatment includes removal of impurities and contaminants from surfaces through the use of an energetic plasma generated from gaseous

**Fig. 4.12**  Cleaning of silicon palettes using boiling acetone

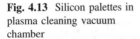

**Fig. 4.13** Silicon palettes in plasma cleaning vacuum chamber

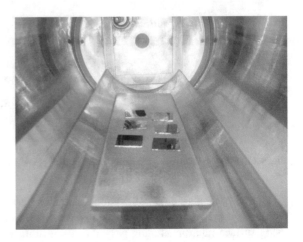

species. Gases such as argon and oxygen, and mixtures such as air and hydrogen/nitrogen are used.

Figure 4.13 represents plasma cleaning equipment chamber with specimens used for the process. The parameters and main constituents of interest of this process were as follows: pressure in vacuum chamber $5 \times 10^{-2}$ torr (here 1 torr = 133 Pa), high frequency generator which produces frequency 13.56 MHz, relative power 0.3 W/cm$^2$, the gas we used was oxygen, aluminum palette to put our specimens was used in order to avoid oxidation, voltage was 1.5 kV and lastly the time of the process was 240–300 s.

As it was mentioned previously having all the specimens cleaned, i.e. with least possible amount of impurities and dirt the deposition of silicon dioxide and polysilicon on wafers took place. Here the procedure was longer more complex and solemn than the ones introduced before. The equipment and sequence used for this procedure is presented in Fig. 4.14.

Having done all the steps mentioned above pallets was nicely covered with uniform polysilicon layer (Fig. 4.15). Nevertheless, not all prepared surfaces were smoothly covered, since dust present in micro fabrication process may have influenced the results.

### 4.3.3   Formation of Micro Membranes

Although deposition of silicon dioxide and polysilicon is clear it comes the most difficult part of the micro-fabrication. Here the important question should be answered: What techniques should be used in order to form micro-membranes of desired thickness and area? According to the tasks generated the area to thickness ratio of membranes is quite big, thus it could lead to fragile and vulnerable

| | |
|---|---|
| Silicon palettes are fixed on a tray | The tray with silicon palettes is fixed to the rotating disk |
| Polysilicon is being deposited on palettes using help of tungsten evaporizer (heated) | Lastly prepared specimens are closed inside vacuum chamber and the deposition process starts after 4 hours when needed vacuum is reached, i.e. 5x10-2 mBa |

**Fig. 4.14**  Equipment used for the polysilicon deposition

**Fig. 4.15**  Smooth mirror
finish polysilicon layer

structures. Nevertheless, photolithography, photoresist application and various etching processes were used in order to obtain final micro sensing element.

### 4.3.3.1   Photolitography

Photolithography is a process used in microfabrication to selectively remove parts of a thin film (or basic weight of the substrate). It uses light to transfer a geometric pattern from a photography chemical photosensitive (photoresist, or simply the "resist") mask on the substrate. A number of chemical treatments then are used to engrave the exposure pattern into the material underneath the photoresist.

Usually and in particular case the process of photolithography starts by covering our already polysilicon covered specimens with photo resist.

Speaking about particular case Shipley Microposit S1805 photoresist, delivered by Rohm and Hass Electronic Materials Europe, was used. The composition and main chemical characteristics of it are presented in Table 4.13.

In order to be more informative and thorough photolithography process is divided in 4 sub processes, namely: photoresist application, exposure, development, etching, photoresist removal and present those using visual aids (pictures) in sequential order.

First of all, photoresist application takes place. Therefore, Fig. 4.16 with pictures and clear explanation of photoresist application is presented. The pictures are

**Table 4.13** Composition and main chemical characteristics of photoresist used

| Composition | Cas-No. | Einecs. No | Concentration | Classification |
|---|---|---|---|---|
| 2-Methoxy-1-methyletyl acetate | 108-65-6 | 203-603-9 | 80.0–<90% | XI, R10, R36 |
| 2-Methoxypropyl acetate | 70657-70-4 | 274-724-2 | 0.25–<0.5% | T R61, R10, R37 |
| Cresol | 1319-77-3 | 215-293-2 | 0.1–<0.2% | T R24/25, R34 |

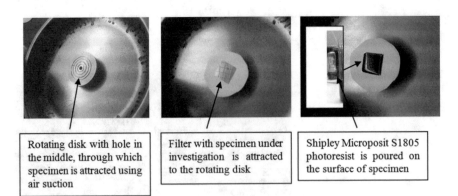

| Rotating disk with hole in the middle, through which specimen is attracted using air suction | Filter with specimen under investigation is attracted to the rotating disk | Shipley Microposit S1805 photoresist is poured on the surface of specimen |

**Fig. 4.16** Photoresist application process

presented correspondingly to the steps of the process. The substrate is coated with photoresist by spin coating. Viscous liquid photoresist solution is dispensed onto the plate and the plate was centrifuged quickly to obtain a uniformly thick layer of photoresist. The spin coating ran at 4200 rpm for 22 s, and produced a layer of photoresist 1.5 μm thick.

### 4.3.3.2   Soft Baking

The photoresist-coated wafer was "soft-baked" to drive off excess solvent (in order to obtain more viscous photoresist). Soft baking was accomplished by infrared rays drying/heating furnace "LADA" (Fig. 4.17). The temperature was gradually increased till 100 °C. The process took 7 min.

### 4.3.3.3   Photoresist Application and Soft Baking (II Time)

Photoresist application and soft baking were performed one more time (II time) having exactly the same parameters, but with photosensitive material to be deposited on another side of the wafer. Soft baking was carried with corresponding specimen side upwards too.

### 4.3.3.4   Mask Alignment and Exposure

After the wafer was baked softly for the second time superpositioning and exposure process was executed. It was achieved by the means of mask alignment and exposure system "JUB 76G" (see Fig. 4.18). The top side of the wafer was exposed by UV light (λ = 365 nm) for about 17 s.

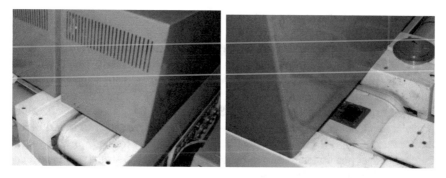

**Fig. 4.17** "LADA" desiccation system: *left-inlet part*; *right-outlet part* (with wafer coming out)

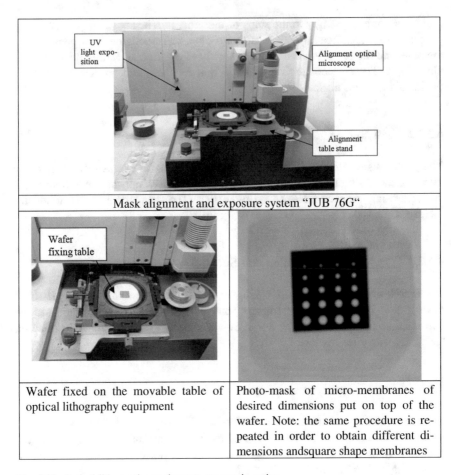

**Fig. 4.18** Optical lithography equipment, process in action

### 4.3.3.5  Thermal Hardening and Si Etching

Following theory, used photoresist after exposure became chemically more stable. This chemical change allowed removing the parts of photoresist, which were not covered by photo-mask. The photoresist had to be removed with a special solution, called "developer" by analogy with the developer. Here the "developer" used consisted of clean dionized water and 0.8% of sodium hydroxide (NaOH), which formed a strong alkaline solution when dissolved in a solvent such as water. Before developer was applied post-exposure "hard-bake" was performed, to help reduce standing wave phenomena caused by constructive and destructive interference patterns of the incident light to secure the remaining photoresist and make more durable protective layer in the future. This was done using equipment shown in Fig. 4.19. The process took 22 min and the temperature was 120 °C.

**Fig. 4.19** Hard-baking equipment

After having done the second sub process (exposure and developing) used in photolithography etching of Si layer took place. For anisotropic wet etching to be executed, tetramethylammonium hydroxide (TMAH) was used. This composition (molecular formula—$(CH_3)_4NOH$) is liquid ammonium salt capable of etching single-crystal Si but leaving $SiO_2$ layer undamaged. It is particularly important that the TMAH (as well as other anisotropic etchants) provide a unique form of Si during its bulk micromachining. In our case the wafer was placed into special quartz vessel containing a bit of concentrated (25%) TMAH. At last the transparent vessel was heated using a hot plate up to 85 °C. The total Si dissolution time was approximately 18 h.

Despite the fact that all theoretical rules were followed very precisely and thoroughly during etching process one of wafers was hardly damaged (see Fig. 4.20). The possible reasons for this, in practice well known as peeling, might have been: some contaminations left during cleaning process of silicon palette, some shortcomings during deposition process, bad timing of etching process or too much pressure exerted while drying the wafer with $N_2$ gun.

### 4.3.3.6   SiO₂ Layer Etching and Photoresist Removal

In this stage reactive ion etching (RIE) technology was used. In physical stage strong radio frequency of 13.56 MHz initiated by generator is applied to the wafer platter (cathode) and provides electromagnetic field between electrodes. The oscillating field ionizes gas molecules by making them get rid of electrons, and in such a way plasma is created. Within each field, the electrons are accelerated electrically cycle up and down in the chamber and sometimes strike the wall of the chamber and the electrode. When the electrons are absorbed in the walls of the chamber, they simply serve to the ground and do not change the electronic state of the system. However, electrons are absorbed into the cathode, causing the platter to

**Fig. 4.20**  Peeling of thin polysilicon layer

create a charge which is developing a large negative voltage (self-bias voltage) due to its DC isolation. Significant voltage difference leads to migration of positive ions towards negative cathode. The specimen placed on the platter is bombarded by very reactive ions resulting in the etching of necessary areas of the wafer. If the ions have sufficient amount of energy, they can sputter atoms out of material without chemical reaction. Process of reactive ion etching is shown in Fig. 4.21.

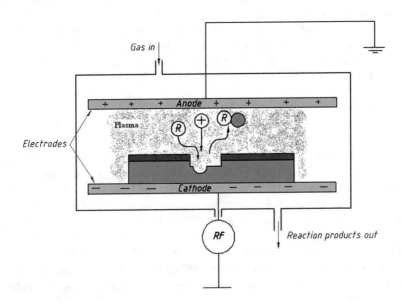

**Fig. 4.21**  Reactive ion etching process [16]

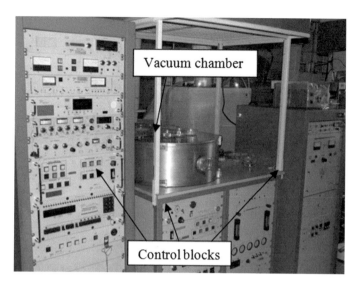

**Fig. 4.22**  RIE equipment PK-2420RIE

In order to obtain freely hanging micro-membranes of required thickness and just of polysilicon above described reactive ion etching of remaining silicon dioxide layer was applied. Here "PK-2420RI" equipment (Fig. 4.22) was used, which is capable of achieving reactions necessary to produce a constant physical and chemical etching environment in the immediate vicinity of the wafer surface. Dry etching techniques can be implemented for many different process steps when manufacturing semiconductor devices. System controlled parameters include plasma watt density, voltage potentials, fixture temperature, process gas flow and reactor pressure.

Basic data of the system is given below.

Base vacuum in chamber: $10^{-3}$ Pa, Etching pressure of microstructures: 0.13 ÷ 70 Pa, RF generator power: 0 ÷ 3 kW, Frequency: 13.56 MHz, Bias voltage of electrode: 0 ÷ 800 V, Temperature of substrate: 15 ÷ 50 °C, Gas: $SF_6$, $N_2$, $O_2$, Ar, He.

The simplified structural scheme of the system is shown in Fig. 4.23 [17]. Note that supplied gases are shown for general circumstances. In specific case only $SF_6$, $O_2$ (for etching) and Ar (for system cleaning) were used.

### 4.3.3.7   RIE Etching Procedure

Constant mixture stream of oxygen and sulfur hexafluoride (10 $cm^3$ of $SF_6$ and 2 $cm^3$ of $O_2$) helped to generate a plasma power that varied from 0.5 to 2.0 kW (power density of the plasma (N), ranged from 0.25 to 1.0 $W/cm^2$) with a pressure (pch) in the cell being 40 Pa. Cathode. Bias voltage ($U_b$), ranged from 200 to

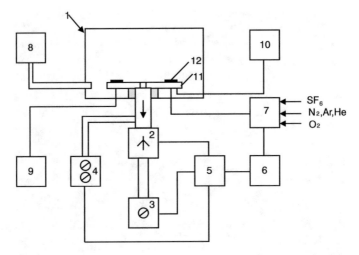

**Fig. 4.23** Simplified structural scheme of PK-2420RIE. Parts: *1*—vacuum chamber, *2*—diffusional pump, *3*—mechanical pump, *4*—double-stage mechanical pump, *5*—control block, *6*—automatic control, *7*—gas flow regulator, *8*—vacuum measurement and regulation block, *9*—electrode temperature measurement and regulation block, *10*—high frequency generator (13,56 MHz), *11*—vacuum chamber electrode, *12*—wafers

300 V. So as to get this a strong radio frequency (f = 13.56 MHz) was initiated by generator resulting in bias voltage of cathode (Up) being 200 or 300 V. Temperature ($T_{body}$) of the wafer was constant (20 °C). The specimen was kept in the chamber for about 10 min. The silicon dioxide was etched by liquid mordant (solution of $NH_4F:HF:H_2O$). It took approximately 65 min for the etchant to reach polysilicon thin film and stop. Once the open $SiO_2$ areas were vanished photoresist remover "mr-Rem 660" was applied submerging the specimen into the vessel with remover for 3 min only.

### 4.3.4   Results of Fabrication, Micro Hardness and Surface Morphology Tests

#### 4.3.4.1   Results of Fabrication

In order to determine if the fabrication process has been successful several individual photographs of micro-membrane were performed using a scanning electron microscope. Analyzing images presented below, it can be observed that the manufacturing process has been specifically unsuccessful with the largest dimension membranes. Figure 4.24 represents fabrication cracks of micro-fabricated micro-membranes. Referring to the theoretical and practical knowledge, most likely causes of the micro-membrane failure and fracture can be:

**Fig. 4.24** Fabrication cracks

(1) residual stresses are too great;
(2) a quantity of dust during the manufacturing process appeared on the surface;
(3) etchant concentration was too high leaving the structure is extremely thin and vulnerable.

Formed membrane of 1 mm side lengths is presented in Fig. 4.25. Comparing with previous case this one provides a view of almost suitably fabricated microstructure. Etched surface is quite uniform and even. Nevertheless, several inaccuracies can be seen too. Firstly, various dots, small holes and valleys are present in membrane surface. This can be explained referring to possible micro-sized particles deposited on membrane and so acting as contaminants in the manufacturing stage. Secondly, edges of top polysilicon are not absolutely straight at mask opening boundaries. Conversely, corrugated shape of the film is seen so underetching phenomenon is not avoided completely.

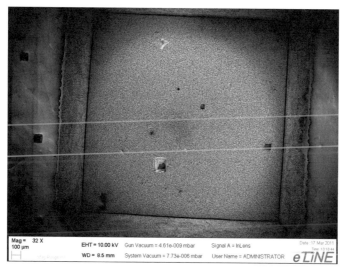

**Fig. 4.25** Membrane with 1 mm side lengths

**Fig. 4.26** Membrane with 0.4 mm side lengths

The most successful fabrication process result is shown in Fig. 4.26. Smooth surface and side walls, clearly seen edges, angles and intersections between surfaces. In addition, no contamination or unnecessary particles, which could enhance distortion or deformation of thin polysilicon are observed. On the other hand, already described corrugated edges of top polysilicon are valid for this case as well.

### 4.3.4.2   Micro Hardness Test

Microhardness term widely used to describe the testing hardness of materials with low loads. A more accurate term is "test Microindentation hardness." When tested for microindentation hardness, diamond indenter impression on the specific geometry of the test surface of the sample using the known applied force (typically referred to as "load" or "load test") from 1 to 1000 gf (gram force) was identified. Microindentation tests usually have 1–2 mN force and produce a pair of recesses microns. Due to their specificity, the microhardness testing can be used to monitor changes in the hardness at the microscopic level. Unfortunately, it is difficult to standardize the measurement of micro-hardness; it was found that virtually any material microhardness is higher than its macrohardness [18]. In addition, the microhardness changes with the load and the hardening material effects. Three of the most commonly used microhardness tests, are tests which can also be applied with heavier load as macroindentation tests: Vickers hardness (HV), Test Knoop Hardness (HK), Martens Hardness (HM).

Computer-controlled measurement system FISCHERSCOPE® HM2000 (see Fig. 4.27) was used for microhardness testing and determining the polysilicon thin

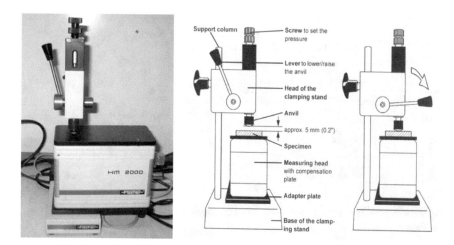

**Fig. 4.27** Microhardness test: *left* microhardness measuring equipment Fisherscope HM2000; *right* principle scheme

film material in accordance with the parameters of ISO 14577. In general, the Martens hardness (HM) is measured with FISCHERSCOPE® HM2000, and it is determined from the area of the indenter displacement under load. The depth of the indentation and a constant, specific to each indenter are used for calculating the displacement of the indenter area.

Advantage Martens hardness measurement as compared with the conventional Vickers measurement is that it is free of subjective factors, such as optical measurement indentation diagonal. Nevertheless, FISCHERSCOPE® HM2000 is also suitable for measuring the hardness or using Berkovich or a spherical indentor.

The procedure of measurement is that specimen is placed directly on a measuring head and is held to the measuring head with an anvil. The indenter, the load generator unit and the indenter displacement measuring system for measuring the indentation depth are incorporated in the measuring head of the device. The standard and default indenter is a diamond pyramid according to Vickers with a 136° face angle. When executing a measurement, the indenter is placed softly and slowly on the specimen ($v \leq$ μm/s) to avoid impacting. The gradually applied load in particular case was up to 1 mN with the step of 0.1 mN. The indenter displacement measurement system for determining the indentation depth has a resolution in the nanometer range.

Micro-hardness results of 0.4 and 1 mm side lengths micro-membranes are presented in the following graphs (see Fig. 4.28). Examining the results and performing additional calculations it can be concluded that micro-hardness value is strongly acceptable, because it differs slightly from a pure silicon.

The average numerical value of the Martens hardness of fabricated objects was HM = 12,543.83 N/mm$^2$ and HM = 10,458.85 N/mm$^2$ for 0.4 mm and 1 mm side length membranes respectively. Invoking theoretical knowledge Martens hardness

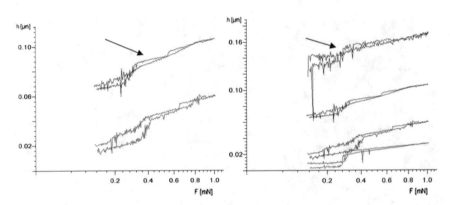

**Fig. 4.28** Micro test results: *left* 0.4 mm side length membrane; *right* 1 mm side length membrane

of pure silicon is 13,476.89 N/mm$^2$ under 1 mN load. Comparing the results it is obvious that fabricated micro-membranes, especially the smallest ones, are hard enough for further applications and experiments.

### 4.3.4.3 Surface Morphology of Micro-membrane

As far as fabricated micro objects will be used for optical applications as basic sensing elements or element matrices as it was mentioned before the surface of them should be as smooth as possible to have perfect reflective characteristics. Here atomic force microscope was employed in order to measure and analyse substance surface characteristics (surface free energy, surface morphology, surface adhesion force, surface roughness). Surface mean height, average roughness, root mean square roughness, valley depth, peak height, peak-valley height, skewness, was measured in order to evaluate various microirregularities of surfaces of interest [18–20].

Atomic force microscope Nanosurf EasyScan 2 (Fig. 4.29) [21] was used to investigate manufactured micro-membranes. The device consists of scanner, controller and connection cables. Scanner is designed for work in open air, and it gives convenient access for sample installation and change of scanning probe. Unlike in most AFMs, where piezotubes are used to move the sample, this scanner is based on electromagnetic movable measuring structure having installed probe, video camera and being supported by 3 levelling screws. Controller serves as a link between the host computer and the scanning unit. Basically it indicates the status of Z-feedback loop and shows it by means of certain colour flashing lights. In addition, software "Nanosurf Easyscan 2" was applied for data processing, visualization and analysis. All measurements were performed in dynamic mode (scanning resolution: 256 × 256, the oscillation frequency of the sensor: 166 kHz, the amplitude of the probe oscillations: 0.2 V, applied probe: NCLR).

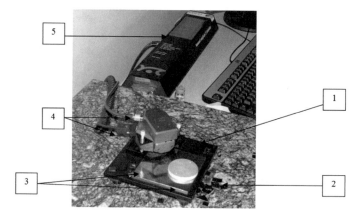

**Fig. 4.29** AFM Nanosurf easyScan 2 Parts: *1*—scanner, *2*—sample holder, *3*—sample stage, *4*—scan head and video camera cables, *5*—controller

Despite the fact that biggest dimension micro object was hardly damaged, there was scientifically important to find out why such phenomenon occurred. Figure 4.30 represents surface morphology of damaged micro-membrane.

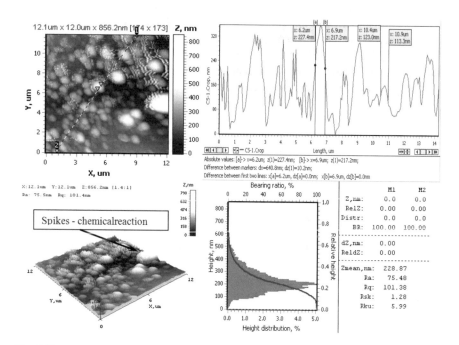

**Fig. 4.30** Surface morphology of damaged micro-membrane: *top-left* measurement of spikes; *top-right* minimal and maximal spike dimensions; *bottom-left*; 3D view of micro-membrane surface; *bottom-right* histogram and surface roughness

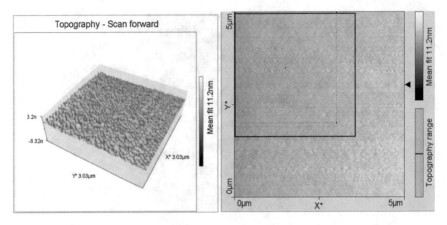

**Fig. 4.31** Surface morphology of 0.16 mm² area membrane: *left* 3D view of micro-membrane surface; *right* area of investigation. Surface roughness parameters: average roughness, nm ($R_a$)—0.451, root mean square roughness, nm ($R_q$)—0.568, peak-valley height, nm ($R_y$)—4.158

Analyzing performed test results and surface morphology graphs it can be assumed that fabrication process was unsuccessful, because of etchant concentration level, causing chemical reaction which can be seen in 3D view of micro object. Or, most common area to thickness ratio problem appeared, which led to critical high thermal stresses distribution level on thin polysilicon membrane surface.

Nevertheless, in order to continue experiment and to adapt fabricated micro object to optical sensor it was clear that only the smallest, i.e. 0.16 mm² area membranes can be used. Thus, AFM analysis of them was performed as well to find out needed surface morphology characteristics (Fig. 4.31).

### 4.3.5   Radial Pulse Analysis Through Application of Fabricated Micro-objects

Introduction of micro-sensors and signal processes into radial pulse analysis, would allow elimination of human error possibility in oriental medicine practice of pulse recognition. A wrist-mountable, watch like device is mounted to maintain the sensor position significantly above radial artery. The pulse waveforms are recorded employing an analog-to-digital converter and a personal computer. To investigate the valuability of performed experiments and obtained pulse forms, it was necessary to examine radial pulse medical characteristic theory in greater detail.

#### 4.3.5.1   In-depth Characteristics of Radial Pulse

The use of radial pulse signal, has aided contemporary medicine in examination of systolic-diastolic blood pressure, stiffness of the arteries and heart rate. These are essential parameters to treatment of subjects with conditions, such as hypertension or atherosclerosis. Utilization of pulsations, recorded at three separate pulse points is not limited to assessment of heart function, but also for the body organs like stomach, small and large intestine, bladder, kidney, liver, spleen and gall bladder. Illustration below (Fig. 4.32) [8] identifies the pulse points as Proximal, Middle and Distal (P, D, M respectively) pulse points, with their relative location from the heart. Generally, a practitioner would observe and analyze the palpations using fingertips. Such approach is highly subjective and is highly dependant on the experience and competence of the practitioner.

Ventricular ejection from heart and subsequent forward wave generation in the arterial structure are the basic phenomena behind pulse signal establishment. Arterial channel possesses several branches which reach completion at the periphery, in the form of several organs. Overall radial pulse morphology is established by a combination of the forward pulse wave and it's reflection from various peripheral organs like kidney, small intestine, large intestine, stomach etc., [4, 5]. In addition, the reflected pressure waves tend to play a major role in determining the patterns of wrist-pulse waveform, by increasing the load to the heart. Pulse wave velocity is identified by the velocity at which the reflected wave travels. Key factors establishing the overall arterial pulse morphology are the nature of wave reflection from peripheral organs and the pulse wave velocity.

From general medical knowledge, it is known that blood pulse comes in the form of a normal wave with resemblance of a sine or a bell curve, located between the organ and the energy (qi) depths. Different qualities such as the hollow full-overflowing, flooding-excess, and flooding-deficient qualities, the suppressed quality, and the hesitant quality result in unusual wave forms. The hollow full overflowing quality exceeds the qi depth, yet is of the correct shape (see Fig. 4.33)

**Fig. 4.32**  Traditional approach of radial pulse analysis [8]

P - Proximal Pulse Point
M - Middle Pulse Point
D - Distal Pulse Point

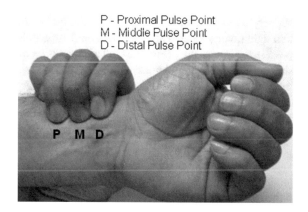

P   M   D

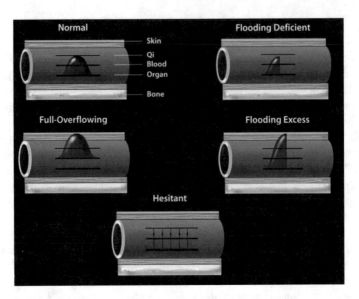

**Fig. 4.33** Qualities of radial pulse waveforms [22]

[22]. Assessment is performed on each side separately when identifying uniform qualities in large segments.

Tense, tight, yielding, spreading, diffuse, reduced substance, vibration, deep, feeble, hollow, hollow full-overflowing, slippery, change in intensity, and cotton are uniform qualities most common to the left side. If the "nervous system" affects the "organ system", the most superficial aspect of the pulse features a slightly feeble deep quality with a thing right quality. When the etiology is persistent, moderate worry, a smooth, superficial vibration can be observed over the entire left side. Parenchymal damage to vital organs manifests itself through rough vibration limited to the left side only.

Tense, tight, yielding, reduced substance, deep, feeble, and cotton are uniform qualities most common to the right side. Also, but much less often a uniform slippery quality can be encountered. A too quick consumption of food results in a thin tight quality located at the pulse's most superficial aspect. Generally the system possesses eight depths (Fig. 4.34). The above-the-qi depth is located between the skin and the qi depth; additionally there is the qi depth and organ depth possessing three additional sub-divisions: bone depth, just above the bone and between the organ depth. While the qi and blood depths include contributions of a particular organ to the qi and blood of the entire organism, the organ depth possesses three more subdivisions of qi, blood and organ that inform of respective states of qi, blood and parenchyma of that organ [22].

Tangibly, the most substantial part of the pulse is the organ depth. The pulse tends to lose width and substance as the pressure gets released on surface-bound pulse. The level of sensation, from light to heavy is directly related to the closeness

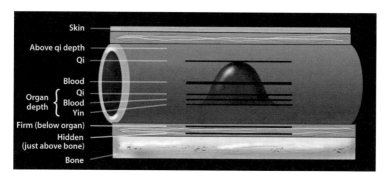

**Fig. 4.34** The eight depths [22]

to the qi depth (for light sensation) and to the organ depth (for heavy sensation). Verbal description of the exact positions is insufficient as the positions are very subtle and can only be conveyed through direct demonstration. The increments between the depths are on the order of tenths of a millimeter, and can only be accessed by observing feedback to incremental changes in pressure (usually applied through practitioner's fingertips). The movement is fully confined to the wrists. The distance from the skin depth to the qi depth when compared to the distances between qi to blood and blood to organ depths, is a about a third of the distance greater. The location of the qi depth is a precise point below the surface, while the organ depth is located at another precise point a significant distance from the bone depth. Located half way between the two is the blood depth. To sum up, the precision of practitioner's wrist is the defining factor in locating each depth and not simply the location of initially encountered vessels by the practitioner.

Even though deeper and more superficial sensations can be discerned at these positions, they hold no correspondence to the qi, blood and organ depths. Quality or intensity variations influence all the depths over the entire pulse or at an individual position.

This particular thesis puts main stress on the modelling, experimental and fabrication activities of computerized noninvasive blood pulse analysis system. Nevertheless, Table 4.14 presents information from the point of view of medicine, regarding types of pulses and their meaning, for comparison to obtained experimental data.

From stand point of mathematical functions, the wrist-pulse curves are in close resemblance to Gamma density function. Wrist-pulse waveforms can be represented as the sum of forward and backward waves, without losing generality, as the following function [24]:

$$
\begin{aligned}
F(t|\alpha, \beta, \Delta, A_w, B) &= A_w f(t|\alpha, \beta, 0) + B f(t|\alpha, \beta, \Delta) \\
&= A_w t^\alpha e^{-\beta t/10} + B t^\alpha e^{-\beta(t-\Delta)/10},
\end{aligned}
\tag{4.12}
$$

**Table 4.14** Characteristics of pulse types [23]

| Pulse type | Interpretation | Comments |
|---|---|---|
| Intermittent pulse | A slow pulse pausing at regular intervals, often occurring in exhaustion of organs, severe trauma, or being seized by terror | As with the scattered pulse, this pulse type is usually only seen in cases where the person is hospitalized or otherwise in an advanced disease stage. It is expected to occur, for example, with those having serious heart disease |
| |  | |
| Fine pulse or Thready pulse | A pulse felt like a fine thread, but always distinctly perceptible, indicating deficiency of qi and blood or other deficiency states | Although the deficiency can be easily detected by other means, some patients can show an artificially robust exterior appearance, while having notable deficiency. Essence deficiency, the result of chronic illness, can give rise to this pulse type |
| |  | |
| Choppy pulse | A pulse coming and going choppily with small, fine, slow, joggling tempo, indicating sluggish blood circulation due to deficiency of blood or stagnation of qi and blood | This has a more irregular pattern than the knotted pulse that also shows stagnation of qi and blood. The severity of the blood disorder is greater. It vibrates and irregularly moves forward, yielding a choppy sensation with brief hesitations or interruptions in movement |

(continued)

**Table 4.14** (continued)

| Pulse type | Interpretation | Comments |
|---|---|---|
| Slippery pulse | A pulse like beads rolling on a plate, found in patients with phlegm-damp or food stagnation, and also in normal persons. A slippery and rapid pulse may indicate pregnancy | While use of the pulse to indicate pregnancy is no longer of value (as more reliable tests are readily available), and while this pulse, like the long pulse is often normal (occurring especially in persons who are somewhat heavy), it is a good confirmation of a diagnosis of phlegm-damp accumulation. It is sometimes referred to as a "smooth pulse" |
| Stringy pulse or Wiry pulse | A pulse that feels straight and long, like a musical instrument string, usually occurring in liver and gallbladder disorders or severe pain | This is similar to the tense pulse, but longer and more tremulous. While severe pain can be easily reported, the wiry pulse confirms the liver and/or gallbladder as the focal point of the internal disharmony |

(continued)

**Table 4.14** (continued)

| Pulse type | Interpretation | Comments |
|---|---|---|
| Soggy pulse | A superficial, thin, and soft pulse which can be felt on light touch like a thread floating on water, but grows faint on hard pressuring, indicating deficiency conditions or damp retention | This pulse is similar to the fine and weak pulses. The pulse sensation felt on light touch gives the impression of being easily moved, as if floating on water; hence, it tends to indicate spleen-qi deficiency with accumulation of dampness. It is sometimes referred to as the "soft pulse" |
| | | |
| Rapid pulse | A pulse with increased frequency (more than 90 beats per minute), usually indicating the presence of heat | The rapid pulse is quite a bit more rapid than a normal pulse, and usually occurs only when there is a serious illness and mainly when there is a fever. The pulse can become rapid from activity prior to pulse taking |
| | | |
| Slow pulse | A pulse with reduced frequency (less than 60 beats per minute), usually indicating endogenous cold | A slow pulse may also indicate a person at rest who normally has a high level of physical activity, so must be interpreted in light of other diagnostic information |
| | | |

Here, Gamma density function is used as the base function, which is made up by the product of one power and one exponential function:

$$f(t|\alpha, \beta, \Delta) = t^\alpha e^{-\beta(t-\Delta)/10},\tag{4.13}$$

In this case $t \geq \Delta$, $f(t|\alpha, \beta, 0)$ is the function of the forward wave and backward wave—$f(t|\alpha, \beta, \Delta)$, $\alpha$ and $\beta$ are parameters of the shape and rate respectively. $A_w$ and B are forward and backward wave amplitudes, respectively, $\Delta$ delay in time between backward and forward waves. According to estimations, such parameters should provide the best correspondence between the pulse waveforms from Oriental medicine and the collective Gamma density functions.

### 4.3.5.2 Experimental Setup

Outlined below are the main components introduce for assembly of the experimental stand. For simulation of heart pulsation and blood flow in radial artery, a simplified closed loop system with a valve for input and a pressurized chamber for output (to imitate variable fluid throughput) were installed. Controller of flow speed enabled control of fluid speed, from 50 to 600 ml/h. Different pulse waves could be produced as a result of possession of different velocity and resultant pressure values. Presented below are the closed loop system and the flow speed controller (Figs. 4.35 and 4.36 respectively).

The requirements for artificial vascular graft are not limited to just being a flexible elastic tube. A U.S. company GORE supplied a specifically produced vascular graft. The hybrid vascular graft from GORE consists of expanded poly-tetrafluoroethylene (ePTFE) prosthesis, with a reinforced nitinol section. The latter, possesses a partial constrain for purposes of easier insertion and deployment in the body. The graft features a continuous lumen with the CARMEDA BioActive surface that consists of covalently bonded, stable heparin of porcine origin, with

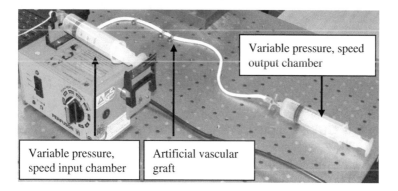

**Fig. 4.35** Closed loop simplified artificial radial artery system

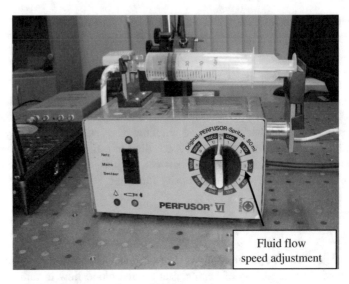

**Fig. 4.36** Fluid flow speed controller used in the experiment

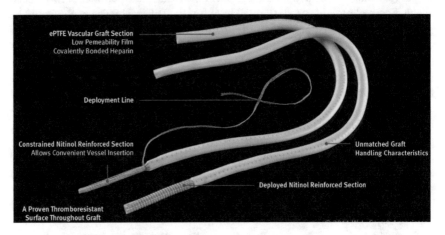

**Fig. 4.37** Vascular graft characteristic [25]

reduced molecular weight. Figure 4.37 and Table 4.15 present main physical and mechanical properties, characteristics of the GORE vascular grafts [25].

Exact material properties of the graft are not described here, as that is proprietary knowledge of GORE company. The prosthesis of choice was meant to imitate the real radial artery, originating from elbow brachial artery and traveling till human wrist, ultimately dividing into smaller arteries and capillaries found in hand, hence, it's length was chosen to be 40 cm.

**Table 4.15** Physical and mechanical properties of expanded polytetrafluoroethylene [25]

| Physical properties | Metric | Mechanical properties | Metric |
|---|---|---|---|
| Density | 0.700–2.30 g/cc | Ball indentation hardness | 27.0–37.0 MPa |
| Apparent bulk density | 0.360–0.950 g/cc | Tensile strength, ultimate | 10.0–45.0 MPa |
| Water absorption | 0.000–0.100% | Tensile strength, yield | 0.862–41.4 MPa |
| pH | 9.5 | Modulus of elasticity | 0.392–2.25 GPa |
| Viscosity | 19.0–25.0 cP | Flexural modulus | 0.490–3.36 GPa |
| | 1.00e+13–1.00e+15 cP | Flexural yield strength | 14.0–27.6 MPa |
| | Temperature 340–380 °C | Compressive yield strength | 1.50–23.4 MPa |
| Deformation | 0.200–14.0% | Shear strength | 9.31–25.5 MPa |

### 4.3.5.3   Experimental Analysis with Artificial Vascular Graft

The computerized MOEMS pulse signal diagnosis procedure, proposed in this thesis consists of three stages: the collection of data, extraction of features and classification of patterns. During the initial stage, micro-fabricated objects attached to the artificial vascular graft of diameter 2.4 mm, are used to collect the pulse signals. To record the displacements of the object, high speed laser triangulation using, high-accuracy KEYENCE LK-G CCD displacement sensor is performed. Presented in Fig. 4.38 are basic structural components and main characteristics of the laser triangulation displacement sensor employed throughout the experiment.

Commonly, laser triangulation sensors consist of a solid-state laser light source and a PSD or CMOS/CCD detector. A part of the laser beam that is projected on the target is reflected onto detector, through focusing optics. The beam moves proportionally with the movement of the target, on the detector (Fig. 4.39) [27]. There are also possibilities, to use laser triangulation sensors on specular surfaces. Typical triangulation sensor, as shown in Fig. 4.39, would not be effective on specular surfaces, due to light being reflected directly, back to the laser. In such cases, the beam needs to be directed towards the target at an angle. The reflected beam will travel from the target at an angle, opposite to the incidence angle and will be focused onto the detector (Fig. 4.39).

The detector signal determines the relative distance from the target. This data is typically available through analog output or digital (binary) and digital display interfaces. CCD and CMOS types of sensors, detect the light distribution at the peak on a sensors array of pixels, to identify the position of the target, while PSD type sensors compute the beam centroid relying on the entire spot reflected on an array.

The analog signal of pulse wave forms, for further examination is obtained during the second stage, using PC Oscilloscope 3000. The oscilloscopes in the PicoScope 3000 series, have sampling rates in real-time, of up to 500 MS/s, bandwidths—of up to 200 MHz and a buffer memory of 128 MS. Due to

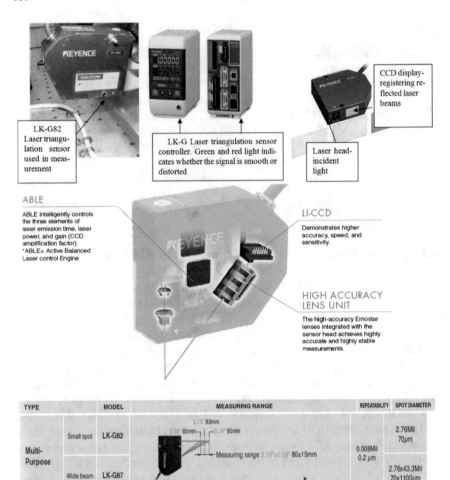

**Fig. 4.38** Laser triangulation displacement sensor used in the experiment, basic structural components and main technical characteristics [26]

availability of a repetitive sampling rate of up to 10 GS/s, detailed display of repetitive pulse signals was possible.

Principal scheme along with the overall experimental model of the initially carried out experiment is presented in Figs. 4.40 and 4.41 respectively. To enable quantitative study of each pulse waveform characteristics, two different pressures were applied to the system—120 mmHg (15,998 Pa, $N/m^2$) and 140 mmHg (18,665 Pa, $N/m^2$). It is apparent, that the first one is meant to represent normal systolic pressure of a healthy person and in the latter case, possibility of hypertension.

Subsequent experiments, focused on variation of fluid viscosity in the system. Artificial blood flow system was introduced to fluids of two different viscosities

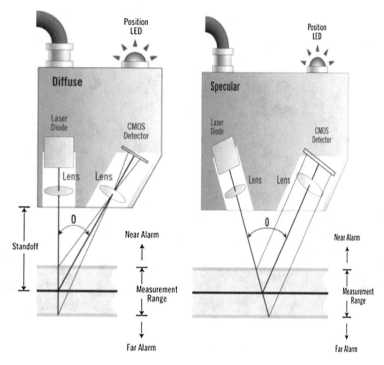

**Fig. 4.39** Laser triangulation sensor principle: *left* diffuse case; *right* specular case [27]

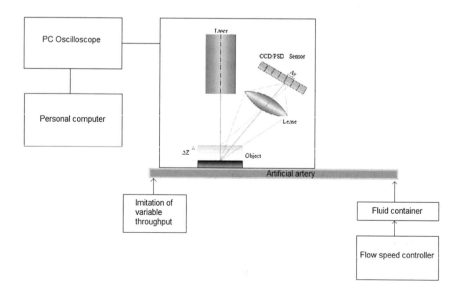

**Fig. 4.40** Principle scheme of the experiment

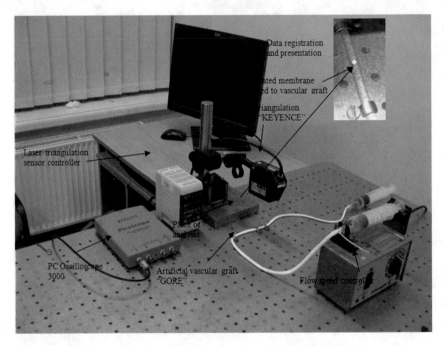

**Fig. 4.41** Experimental model of the artificial blood flow system [28]

$(8.90 \times 10^{-4}$ and $3.2 \times 10^{-3}$ Pa s)—the displacements of investigated points were recorded. Figure 4.42 (top) and 4.42 (bottom) display the recorded data. Presented in Fig. 4.42 (bottom) is the case, when solution with similar specifications to blood was employed. Figure 4.43 demonstrates the case, that employed simple water solution, but under different pressure inputs.

Through analysis of the graphs it becomes apparent, that some misalignments may be encountered when comparing artificial system impulse wave generation with theoretical background. General reasoning for this is the difference in the elasticity modula of artificial graft and human radial arteries. Hence, the resultant displacement values are greater. Regardless, "in vitro" experiments have proven the efficacy of using signal filtering and pattern matching sensitive system for pulse analysis. In addition, it should be noted, that recorded displacement values differ significantly, for different fluid viscosities and input pressures. At fluid viscosity of $8.90 \times 10^{-4}$ Pa s, while under the input pressure of 15,998 Pa the recorded displacements were 31 μm for forward pulse wave and 16 μm for the backward one. With the increase of the pressure value to 18,665 Pa, the displacements were observed to be 39 and 30 μm for forward and backward pulse waves respectively. In fluids with lower viscosity values (e.g. $3.2 \times 10^{-3}$ Pa s) while under pressure of 15,998 Pa, the recorded displacements were significantly smaller, 19 μm for forward pulse wave and 11 μm for backward pulse wave, 28 and 20 μm under input

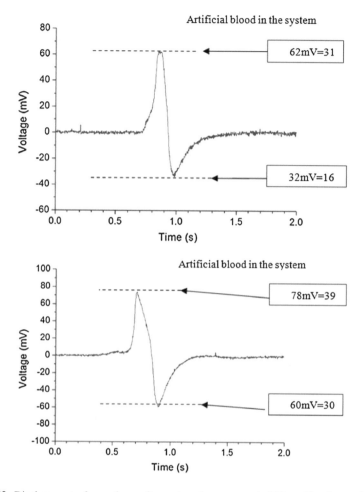

**Fig. 4.42** Displacement of vascular graft: *top* impulse pressure is 120 mmHg; *bottom* impulse pressure is 140 mmHg. Blood like fluid

pressure of 18,665 Pa for forward and backward pulse waves respectively. It was observed, that with the increase in the viscosity of the fluid, the displacements increased as well.

#### 4.3.5.4   Blood Flow Velocity Analysis

Blood flow velocity is to measure the rate at which blood moves through a particular vessel. Examining the blood velocity many characteristics about artery wall clearance, cholesterol levels can be extracted. Just numerous medical experiments should be performed and a lot of medical practice should be gained in order to

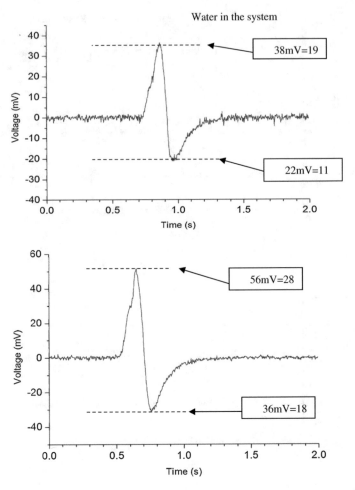

**Fig. 4.43** Displacement of vascular graft: *top* impulse pressure is 120 mmHg; *bottom* impulse pressure is 140 mmHg. Water like fluid

classify illnesses according blood flow velocity. Principle scheme of proposed techniques working principle was presented in Fig. 4.40 [28].

The rate of blood flow in the experiment was chosen to be 600 ml/h. The clamping point was situated 25 cm from the measurement point. Figure 4.44 represents the ideal case, when the pure and perfect artificial graft artificial blood was used in the experiment represents ideal case when clean artificial graft and perfect artificial blood was used in the experiment.

In a particular case, the liquid has traveled in 25 cm 5.2 s. Using the Hagen-Poiseuille equation, which gives a pressure drop in the fluid flowing through the long flexible pipe, it is possible to find the relationship between speed and fluid viscosity. Only equation assumption is that a laminar flow through a tube of

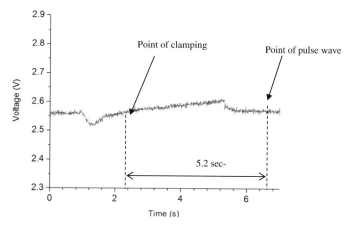

**Fig. 4.44** Registering blood velocity

constant circular cross-section which is substantially greater than its diameter; there is no acceleration and no fluid in the pipe. For a compressible fluid like blood volume flow in a tube, and the linear speed is not constant along the tube. The flow is usually expressed in outlet pressure. Since the fluid is compressed or extended, the work is done, and the liquid is heated and cooled.

This means that the flow rate depends on the heat transfer to and from the fluid. Thus, the volumetric flow rate at the pipe is given by:

$$\Phi = \frac{dV}{dt} = v\pi R^2 = \frac{\pi R^4 (P_i - P_o)}{8\eta L} \times \frac{P_i + P_o}{2P_o} = \frac{\pi R^4}{16\eta L}\left(\frac{P_i^2 - P_o^2}{P_o}\right), \qquad (4.14)$$

wherein $P_i$ the inlet pressure, $P_o$ outlet pressure, L is the length of the tube, $\eta$ is viscosity, R is the radius, V is the volume of fluid at outlet pressure, $v$ is the fluid velocity at the pressure outlet.

### 4.3.5.5 Investigation of Displacements of Vascular Graft by Holographic Interferometry

Holographic interferometry technique is known as live holographic interferometry. The hologram contains all the information on the deformation of the object surface [29, 30]. Temperature conditions may affect the optical path length of the object and reference beams at the time of exposure, and thus affects the phase difference and the quality of the hologram [31, 32].

For visual comparison of experiments executed with artificial vascular grafts, there was a need to check if the obtained results by laser triangulation displacement sensor are valid for further experimentation. Thus, the holographic method was used to analyze the stability of the optical scheme. The tests used the PRISM

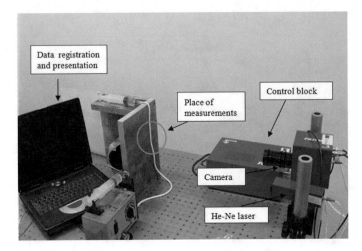

**Fig. 4.45** Holographic interferometry stand used in the experiment

system. Figure 4.45 represents experimental stand used for particular measurements.

Figure 4.46 shows a holographic interference pattern on the fixed surface by adjusting the holder. White areas in the hologram and a small number of interference lines correspond to a very small field of pressure deformation of the vascular graft.

In order to find out dominating displacements of place under investigation displacement spectral analysis is generated. Characteristic graph is obtained using the values of displacements going across the center of prosthesis through all the

**Fig. 4.46** Hologram of vascular graft under pulse wave

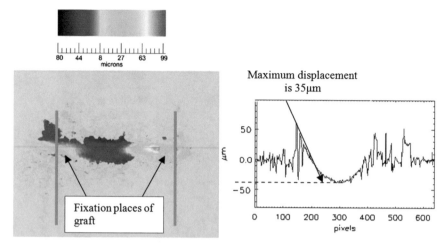

**Fig. 4.47** Displacement measurement of vascular graph under pressure: *left* displacement spectra; *right* characteristic graph

length under investigation. Figure 4.47 represents displacement spectra and characteristic graph respectively.

Comparing obtained displacements using laser triangulation displacement sensor and the ones obtained employing holographic interference techniques it was noticed that maximal displacement values under same conditions (same liquid viscosity and same inlet pressure) differs slightly just about 4 μm. Following the analysis presented above it is observed that using triangulation sensor maximal vascular graft displacement is 31 μm and in holographic interferometry case displacement is 35 μm under the same 15,998 Pa inlet pressure. Such a little difference can be neglected and it can be proved that chosen laser triangulation techniques can be used for further analysis and prototyping for novel MOEMS pulse analysis sensor.

### 4.3.5.6  "In Vivo" Experiments with Patient

Numerous experiments have been carried out with the 28-year-old man. The analysis shows extremely desirable results, since without any pressure applied on the wrist, pulsation of points of interest were recorded. "In vivo" experiment executed is presented in Fig. 4.48.

Figure 4.49 shows the obtained results and characteristical blood pulse peaks obtained executing the experiment.

Analyzing case (Fig. 4.49 right) it is obvious that the obtained pulsation signal in this case has a lot of similarities to the pulse wave characteristics presented in theoretical chapter of the thesis. Forward pulse wave and backward pulse wave can be easily distinguished from the graph. It is worth repeating that the measurements

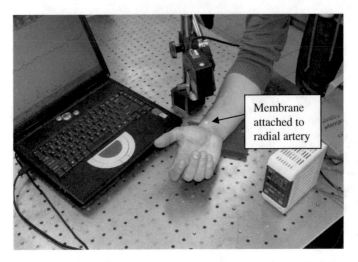

**Fig. 4.48** "In vivo" experiment

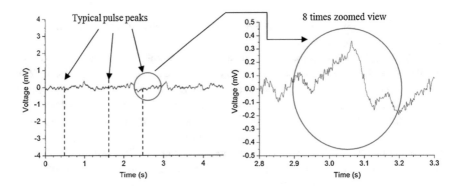

**Fig. 4.49** "In vivo" experimental results of blood pulse: *left* typical pulse peaks in 4 s interval; *right* 8 times zoomed view of pulse wave present in third second

were taken from micro objects, which were freely attached on the pulsation point without any pressure applied on the radial artery. If micro fabricated membranes could reach the radial artery closer (under pressure) the obtained displacement field would be bigger in values and the MOEMS would be more sensitive. Nevertheless, analyzing particular experiment it is evident that filtering from possible noise is needed in order to examine the obtained characteristics more thoroughly and assign them to one of the possible models of traditional oriental medicine pulse patterns.

All the experiments were performed with single micro fabricated membrane and just single point displacements were measured. In order to obtain more sensitive MOEMS there is a need to have matrix of micro membranes, incident laser light splitter, CCD able to read numerous of reflected laser beams and programmed

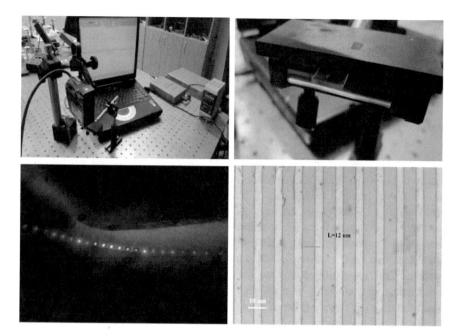

**Fig. 4.50** Proposed system for sensitive MOEMS: *top-left* experimental stand; *top-right* laser beam diffraction grating; *bottom-left* process in action; *bottom-right* diffraction grating structure, period 12 μm

software able to process the signals obtained. Such proposed system is presented in Fig. 4.50.

Here, laser beam splitter (diffraction grating with spacing of 12 μm and slightly trapezoidal shape grooves with a height of 500 nm) was made from a fused silica substrate using conventional optical lithography and reactive ion etching. Reactive ion etching was performed to the quartz substrate $CF_4/O_2$ plasma at 0.6 $W/cm_2$ RF power density using a mask made of aluminum. After etching, the latter was removed in $Cr_2O_3:NH_4F:H_2O$ solution.

## 4.4   Modeling and Simulation of Radial Artery Under Influence of Pulse

Continuing the research there was a need to employ computer numerical analysis in order to simulate the experiment which was executed in the dissertation. Performing modeling it could be checked if the described problem is formulated correctly. If the experimental results coincide with the numerical simulation it could be proved that such programmed numerical analysis is suitable for such type of problem solution. Therefore, treating obtained numerical modeling as basis, considerably more experiments can be done changing radial artery wall thickness, blood like fluid viscosity, inlet pressure, geometry and mechanical parameters of radial artery, etc. For solution of such problem commercial numerical program ANSYS CFX was applied. Complex Computational Fluid Dynamic (CFD) analysis was used.

## 4.4.1   Computational Fluid Dynamics (CFD)

CFD is popular modeling idiom to study fluid flows in variety of applications. This method uses a numerical simulation to solve the mathematical equations involved in the flow of liquid. Usually, such kinds of equations are in the partial differential equations form. Such problems can deal with simple flow characteristics or taking into account complex factors like turbulence and heat flow. CFD has a number of advantages over experimental fluid dynamics (EFD) investigations. The first advantage that CFD has over EFD is in terms of feasibility of design and development. Collecting values of flow data at more than one location requires the construction of a movable sampler data or multiple data-probes arranged so as to cover the area of interest. It is not easy to design conditions when the good sample results can be obtained without subjecting the beyond accuracy preventing measuring devices interfere with the fluid flow [33].

A second advantage that gives the CFD a lead is that extensive simulations at the present time are quite cost-effective compared to the experimental alternative. The programming codes which are used to solve the fluid dynamics equations are commonly available and even basic computers are capable of providing accurate results for complex simulations. Comparatively, the cost required to run the experiments is significantly higher even for simple experiments. That is why the CFD is cost effective and accurate and hence has become a widely-used approach in fluid dynamics calculations. In spite of all advantages of the CFD discussed above, it should be noted that the experimental methods still play a vital role in research. In a CFD technique there will be a small degree of inaccuracy due to round-off errors, which should preferably be kept to an extremely small value.

### 4.4.1.1   Principles of CFD

CFD breaks the geometry in small cells, and therefore, is aimed at solving the basic equations for each cell. These equations describe the physical aspects of fluid flow, and for ordinary flow they are:

Conservation of mass,
Conservation of momentum,
Conservation of energy if heat is transferred from one point in space to another.

For this study, the basic governing equations of flow are the Navier-Stokes equations that determine the conservation of mass and momentum. Particular process of CFD can be divided into three different phases. The first is the pre-processing stage, in which the domain of calculation is determined and it is decomposed into discrete cells to form a mesh or network of cells. A finer mesh will provide the best results, but require more processing power. After the fluid and the equations to be solved have been determined appropriate boundary conditions are applied to the model. This is done by constraining the nodes of cells along the domain boundary with known or controlled parameters. The solution for the rest of the flow field is determined from these boundary values.

In the second stage of CFD, the mesh generated in the previous stage is solved. The solution starts by assigning the cells with an initial value which provides initial estimates for the solution methods. There are several numerical solution techniques like finite difference, finite element and finite volume methods for this purpose. In this work we use methods based on the finite element solvers.

The third stage is post-processing stage which is the final stage in the CFD method in which the data generated from the preceding step is extracted and analyzed.

### 4.4.2 Characteristics of Arteries

Most of the biological tissues generally exhibit an isotropic property, which is they have geometrically uniform dimensions [34–37]. For calculations the assumption is made that tissue of artery is isotropic.

Biological materials in the human body, as the arteries have the character of being deformed in stressful situations. If an artery is exposed to pressure in the form of the mechanical load or compressive force, the deformation occurs and the stresses developed within this arterial wall. The amount of deformation and stress depend on the history and the speed of the applied load, temperature, etc., when the deformation is very small and adiabatic (i.e. no heat is gained or lost), then most likely, that the condition that the stress and strain are independent on loading rate, and that the material is likely to return to its original configuration once the compressive force is removed.

If the deformation is the only stress factor in the elastic body (in this case an artery), it can be described by the Hooke's law.

### 4.4.3 Characteristics of Blood

Blood is formed of erythrocytes commonly known as "Red Blood Cells", leukocytes known as "White Blood Cells", and platelets that are suspended in the plasma. From physical point of view, since blood is a fluid which consists of different

**(a)**

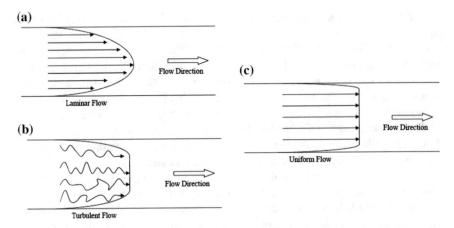

**(b)**

**(c)**

**Fig. 4.51** Flow profiles for laminar (**a**), turbulent (**b**) and uniform (**c**) flow [40]

constituents, it is viscous due to the friction of a two different layers pass over each other. In contrast to simple fluids as water, blood is a suspension of particles, therefore the viscous properties of blood are complex and therefore blood is a non-Newtonian fluid [38, 39]. Viscosity of blood at normal physiological conditions for an average human being is in the range of $3 \times 10^{-3}$ Pa s (Pascal seconds) to $4 \times 10^{-3}$ Pa s (Pascal seconds) with a density of 1060 kg/m$^3$.

A fluid flowing in a vessel exhibits motion which is classified into two types: laminar or turbulent. In laminar flow the fluid particles move along smooth paths in the layers with each layer of sliding smoothly over its neighbor. It is also called streamline flow. As the velocity of flow increases, the different layers start interfering in each other's motion due to different inertia and friction at some point and eventually the laminar flow becomes unstable, vortices start to form, and ultimately the fluid flow becomes turbulent. Hence turbulent flow is one in which different layers collide with one another and cause non-uniformity. When flow is turbulent fluid particle velocity vectors for each point, change rapidly over time, both in magnitude and direction. Laminar and turbulent flows have different designs developed as shown in Fig. 4.51.

For each fluid to achieve turbulent flow, there is a threshold value dependent on the average liquid flow rate, density, viscosity, and vessel diameter (for internal flows). This characteristic quantity that depends on all of these values is called the Reynolds number, as shown in Eq. 4.16.

$$R_e = \frac{\rho u D}{\mu},\qquad(4.15)$$

where u the average velocity of the liquid, D is the diameter of the vessel, $\rho$ the fluid density, $\mu$ is the dynamic viscosity of the fluid. It has been found through experiments that the blood flow tends to be turbulent at a Reynolds number of about 2000 for the internal flow [39, 41, 42].

### 4.4.4 Computational Fluid Structure Interaction (FSI) Modeling for Blood Flow in Radial Artery

This section aims to provide a description of the governing mathematical equations and the algorithm developed for the computational FSI modeling of blood flow in radial artery. Moreover, performing numerical simulation of experiment the mathematical equations used for the solution process and the boundary conditions has also been incorporated. Methods used to attain the solution, as well as a discussion on the FSI coupling method are also included in this subchapter.

#### 4.4.4.1 Blood Flow Modeling

The Navier-Stokes equations are the principal equations to model fluid flow. Under the conditions of an incompressible material, the flow was assumed to be steady so that the time derivative is neglected, $\rho_f g$ denotes the gravitational external force which is assumed to be zero. Thus, general Navier-Stokes equation is simplified to the continuity equation (the conservation of mass) Eq. 4.16 and the conservation of momentum Eq. 4.17 [43].

$$\frac{\partial u_i}{\partial x_i} = 0, \tag{4.16}$$

$$\rho u_j \frac{\partial u_j}{\partial x_j} = -\frac{\partial P}{\partial x_i} + \mu \frac{\partial^2 u_i}{\partial x_j \partial x_j}, \tag{4.17}$$

where $u_i = (u, v, w)$ is the local velocity, $x_i = (x, y, z)$ is the length coordinate, p is the fluid pressure, $\mu$ is the dynamic viscosity. The expression in Eq. 4.18 denotes the surface force $e_{ij}$ and is the strain rate tensor, which generated the stress in the elastic material. The blood like fluid was modeled in isothermal, incompressible and Newtonian (constant viscosity) conditions, and the stress tensor is generated in the blood defined by the constitutive equations is given by:

$$\sigma_{ij} = -p\delta_{ij} + 2\mu e_{ij}, \tag{4.18}$$

where $\sigma_{ij}$ is the stress tensor, and $\mu = 3.3 \times 10^{-3}$ Pa s is the fluid viscosity, and $\delta_{ij}$ is the knockers delta.

### 4.4.4.2   Radial Artery Wall Modeling

The governing equations for the motion of an elastic solid are mathematically described by the following equation [44]:

$$\rho_w \frac{\partial^2 e_{ij}}{\partial t^2} = \frac{\partial \sigma_{ij}}{\partial x_j} + \rho_w F_i, \quad \text{for } i, j = 1, 2, 3, \qquad (4.19)$$

where $e_{ij}$ and $\sigma_{ij}$ are the components of the displacements and stress tensor in a solid respectively, $\rho_w$ is the solid density, $F_i$ are the components of body force acting on solid. Solid material deformation is produced as a result of flow, and according to the law of conservation of energy, the energy will store inside the material.

A single-layered arterial wall was assumed throughout the entire model. The artery was modeled as a hyper-elastic material or Neo-Hookean elastic solid material which shows nonlinear dependence of stress-strain behavior of materials undergoing large deformations [45]. The artery was modeled as an incompressible material since the incompressibility is a reasonable assumption, because biological tissues contain mostly water, which is incompressible at physiologic pressures. The material is incompressible in the sense that its resistance to volume changes in orders of magnitude greater than its resistance to shape changes.

### 4.4.4.3   Modeling Steps

Aim of particular numerical analysis is fluid pulse propagation through flexible vessel (radial artery) and its numerical modeling. Therefore (FSI) problem was solved. As previously mentioned numerical analysis was performed using commercial software ANSYS CFX and the geometry of the structure was completed in ANSYS Workbench environment.

Firstly, artificial radial artery and fluid models were created. For the mechanical, physical and material properties refer to Table 4.6. Here, length of artery was 40 cm; outside radius 2.4 mm and wall thickness 0.35 mm. Geometric parameters correspond to ones used in the experiment. Here, vascular graft is clamped at both ends. Under normal operating conditions arteries already are under pressure and ambient pressure is as well present. Therefore, in order obtain dynamic fluid pulse propagation solution, FSI steady state solution was solved first. Figure 4.52a represents artery finite element model consisting of 1340 elements having 1360 node points and Fig. 4.52b represents model of fluid present in the artery, which consists of 7571 rectangular finite elements having 8432 node points. Creating the model finite elements is fined in the artery border.

For static analysis pressure drop in the system was chosen to be 50 Pa. Figure 4.53 a represents the pressure field of the artery and (b) Von Misses stress distribution on artery wall under internal pressure of 50 Pa.

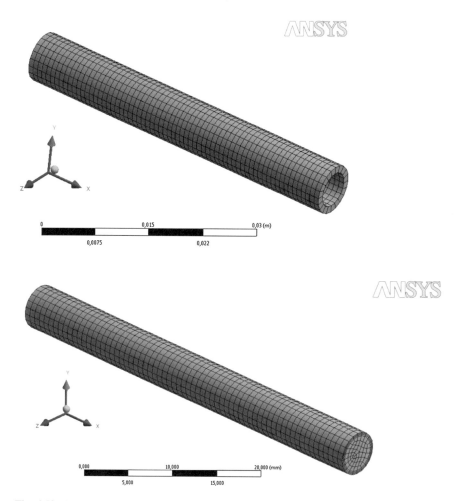

**Fig. 4.52** Geometry of artery: *top* finite element model of artery; *bottom* finite element model of fluid

After solving the steady state FSI problem the artery then was subject to a pressure pulse at the inlet, with peak pressures of 15,998 Pa and duration of 0.5 s (see Fig. 4.54). The transient FSI response of the artery was simulated.

For reference fluid in the system was chosen to be blood like fluid with viscosity $\mu = 3.3 \times 10^{-3}$ Pa s. The emphasize should be made that all the parameters present in the experiment were chosen to be the same in the numerical analysis modeling as well. Displacement registration point was chosen to be 25 cm from the pulse point. Figure 4.55 represent FSI response of the artificial artery.

Comparing the obtained simulation results with experimental data, it can be concluded that such FSI numerical algorithm is suitable for further analysis.

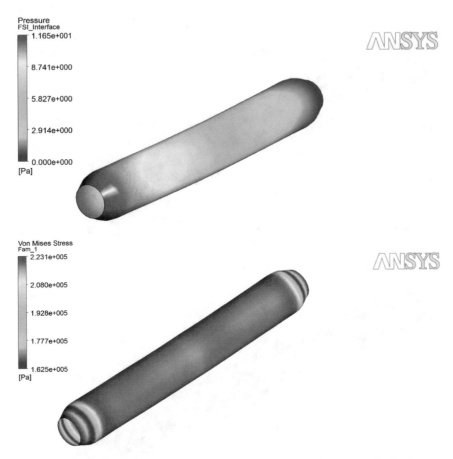

**Fig. 4.53** Static analysis of artery: *top* pressure field; *bottom* von misses stress distribution

Table 4.16 represent the comparison of experimental and modeled displacements of artificial artery under inlet pressure of 15,998 Pa.

Following sequence of experiment, modeling continued with numerical simulation of radial artery with parameters usually present in healthy person (see Table 4.17).

For registering displacements of radial artery points presented in Fig. 4.56 was chosen. Here displacement registration point 1 is situated 3 cm from the inlet pressure point and displacement point 2 is situated 30 cm from the inlet pressure point. The second registration point imitates approximate measuring point present in the "in vivo" experiment.

Displacement graphs of the chosen radial artery points are presented in Fig. 4.57.

Analyzing obtained results it is obvious that there is a slight difference with the results present in experimental case. It can be explained that simulation was done

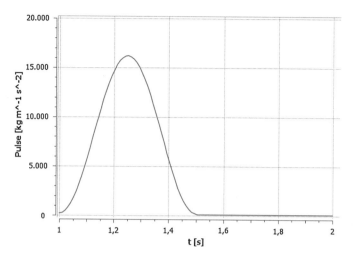

**Fig. 4.54**  Applied impulse pressure (inlet pressure) 15,998 Pa

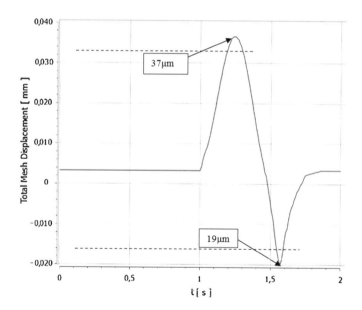

**Fig. 4.55**  Displacements of artificial artery, inlet pressure 15,998 Pa

| **Table 4.16**  Experimental and modeled artificial vascular graft displacements | Inlet pressure 15,998 Pa, Fluid viscosity $3.3 \times 10^{-3}$ Pa s | | |
|---|---|---|---|
| | Experimental displacement results (μm) | 37 | 19 |
| | Simulated displacement results (μm) | 33 | 16 |

**Table 4.17**  Parameters of radial artery [34]

|  | Control normotensive | Hypertensive |
|---|---|---|
| *Parameters at MAP* | | |
| Internal diastolic diameter, mm | 2.53 ± 0.32 | 2.50 ± 0.56 |
| Intima-media thickness, mm | 0.28 ± 0.05 | 0.40 ± 0.06[***] |
| Wall cross-sectional area, mm$^2$ 0.23 ± 0.04 | 2.45 ± 0.57 | 3.79 ± 1.14[***] |
| Intima-media thickness/internalradius ratio | 0.23 ± 0.04 | 0.33 ± 0.08[***] |
| Circumferential stress, kPa | 52.6 ± 11.0 | 50.1 ± 10.5 |
| Distensibility, kPa$^{-1}$ × 10$^{-3}$ | 5.48 ± 4.10 | 5.03 ± 3.52 |
| Cross-sectional compliance, m$^2$ × kPa × 10$^{-8}$ | 2.71 ± 2.09 | 2.42 ± 1.80 |
| Elastic modulus kPa × 10$^3$ | 2.68 ± 1.81 | 2.25 ± 2.14 |
| *Parameters at 100 mm Hg* | | |
| Distensibility, kPa$^{-1}$ × 10$^{-3}$ | 4.21 ± 1.83 | 7.59 ± 6.45[*] |
| Cross-sectional compliance, m$^2$ × kPa$^{-1}$ × 10$^{-8}$ | 2.10 ± 1.55 | 3.46 ± 2.41[*] |
| Elastic modulus, kPa × 10$^3$ | 3.28 ± 2.11 | 1.84 ± 1.65[**] |

MAP indicates mean arterial pressure. Value are mean ± SD
[*]$P < 0.05$; [**]$P < 0.01$; [***]$P < 0.001$

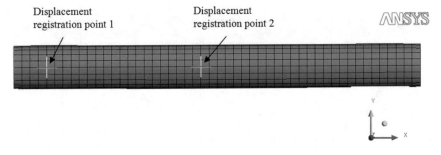

**Fig. 4.56**  Displacement registration points

simplifying the problem disregarding rheology of muscles, skin, etc. Nevertheless, displacement graph of point 2 has similar characteristics to described pulse wave forms presented in previous experimental sections. Therefore, further analysis can be continued.

## 4.5  Moiré Method Application for Artery Surface Deformations Analysis

The moiré projection techniques enable obtaining of the object relief [46–48]. This section uses the classical application of the moiré projection method, which measures the out-of-plane displacements based on the differences between two

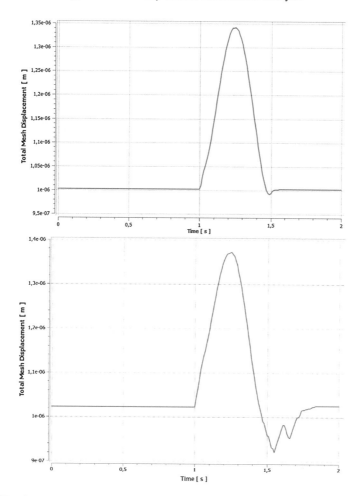

**Fig. 4.57** Displacement of radial artery: *top* displacement of point 1; *bottom* displacement of point 2

prompting states. A fringe pattern can be produced through double exposure in both states, to extract the displacement field.

## 4.5.1   *Mathematical Representation of the Projected Image*

The following steps utilize the paraxial model (approximation of such condition is available through the use of a slide projector and imaging system, placed far away from the specimen). The issue of depth of focus is an important factor,

**Fig. 4.58** One-dimensional geometrical representation of the optical projection on a diffuse surface; F(y) is the projected image, G(x) is a diffuse deformed surface, and H(z) is the observed image [49]

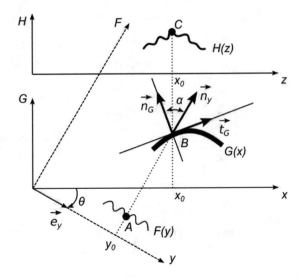

consideration of which is essential to practical applications. The camera focuses on one plane at a time, resulting in coordinate error in out-of-plane points. Use of paraxial model, eliminates the depth of focus issue as well as perspective effects resulting from the use of a point light source.

An assumption is made, that the x-axis is perpendicular to the direction of observation, while θ is the angle between the illumination and observation directions. Provided below (Fig. 4.58) is a one-dimensional representation of the optical projection on a diffuse surface [49].

Defined in the frame y0F is the projected image F(y), the frame is rotated with respect to the frame x0G by an angle θ. The function F(y) defines the level of greyscale of a white light ray that travels through point y0 along the F-axis. Consequently, if $0 \leq F(y) \leq 1$, where 0 is the black colour; 1—the white colour, and all the values in between represent appropriate greyscale levels. A practical example of F(y) is provided in Fig. 4.58 making the development of geometrical relationships less complicated. Equations provided below describe the image construction procedure.

$$H(z) = H(x) = F(y) \cos \alpha; \tag{4.20}$$

$$\cos \alpha = (\vec{n}_y \cdot \vec{n}_G) = \frac{\cos \theta + G'_x(x) \sin \theta}{\sqrt{1 + \left(G'_x(x)\right)^2}}. \tag{4.21}$$

The below equality is valid for all x and y above 0:

$$y = x \cos \theta - G(x) \sin \theta. \tag{4.22}$$

Hence, finally:

$$H(x) = F(x \cos \theta - G(x) \sin \theta) \frac{\cos \theta + G'_x(x) \sin \theta}{\sqrt{1 + \left(G'_x(x)\right)^2}}, \tag{4.23}$$

where F(y) is the image being projected, G(x) is a deformed, diffuse surface, H(z) is the image being observed, θ is the angle between the illumination and observation directions.

## 4.5.2 Double-Exposure Projection Moiré

Double-exposure moiré projection technique consists of two steps. In the beginning, the grating is projected oblique, to the viewing direction of the G(x) surface, with the grating that is observed, being photographed. Subsequently, the specimen gets deformed (the projection and imaging systems are not affected), which is followed by another photograph of the grating. Upon superposition of the obtained images, moiré fringes are produced, which can subsequently be used to identify the magnitude of the specimen's deformation.

The deformed specimen surface is described as G(x) + g(x), here g(x) is the absolute deformation of value in the direction of the observation, after the application of the load. In particular, work an assumption is made, that the image being projected is the harmonic moiré grating. Hence:

$$F(y) = \frac{1}{2} + \frac{1}{2} \cos\left(\frac{2\pi}{\lambda} y\right), \tag{4.24}$$

where the pitch of the grating is equal to λ. Additionally, the function g(x) is a slowly varying function in this particular case:

$$\frac{\cos \theta + \left(G'_x(x) + g'_x(x)\right) \sin \theta}{\sqrt{1 + \left(G'_x(x) + g'_x(x)\right)^2}} \approx \frac{\cos \theta + G'_x(x) \sin \theta}{\sqrt{1 + \left(G'_x(x)\right)^2}}. \tag{4.25}$$

Subsequently, subtractive superposition of the observed grating before and after the load is performed:

$$\frac{1}{2} + \frac{H_1(x) - H_2(x)}{2} \times \frac{\sqrt{1 + (G_x'(x))^2}}{\cos\theta + G_x'(x)\sin\theta}$$

$$\approx \frac{1}{2} + \frac{1}{4}\left(\cos\frac{2\pi}{\lambda}(x\cos\theta - G(x)\sin\theta) - \cos\frac{2\pi}{\lambda}(x\cos\theta - (G(x) + g(x)\sin\theta))\right)$$

$$= \frac{1}{2} - \frac{1}{2}\left(\sin\frac{2\pi}{\lambda}\left(x\cos\theta - G(x)\sin\theta - \frac{g(x)\sin\theta}{2}\right)\right) \times \sin\left(\frac{2\pi}{\lambda}\frac{g(x)\sin\theta}{2}\right)$$

$$(4.26)$$

where $H_1(x)$ and $H_2(x)$ are, gratings observed before the load and after the load. The preceding equation is meant to represent the effect of beats; the envelope function is

$$\frac{1}{2} \pm \frac{1}{2}\sin\left(\frac{2\pi}{\lambda}\frac{g(x)\sin\theta}{2}\right) \tag{4.27}$$

Moiré fringes will be generated at

$$\frac{\pi g(x)\sin\theta}{\lambda} = \pi N; \quad N = 0, \pm 1, \pm 2, \ldots \tag{4.28}$$

Finally, the displacement $g(x)$ in terms of fringe order N reads:

$$g(x) = \frac{(0.5 + N)\lambda}{\sin\theta}. \tag{4.29}$$

### 4.5.3   Two Dimensional Example

Generally, the plane of the image, the plane of the surface and the plane of observation may be located independently in three-dimensional space. An assumption is made, that the angle of observation is equal to zero, the H and G planes are parallel (Fig. 5.60), illumination angle is $\theta$; angle between the F and G planes is also $\theta$. The function F(y, v) is used for determining the greyscale levels of the image being projected; GK (x, v)—the surface shape; HK (x, v)—greyscale levels of the image being observed. Then, Fig. 4.59

$$F(x, y) = \frac{1}{2} + \frac{1}{2}\cos\left(\frac{2\pi(x\cos\alpha - v\sin\alpha)}{\lambda}y\right), \tag{4.30}$$

$$H_K(x, v) = F(x\cos\theta - G(x, v)\sin\theta, v)\frac{\cos\theta + \frac{\partial G_K(x,v)}{\partial x}\sin\theta}{\sqrt{1 + \left(\frac{\partial G_K(x,v)}{\partial x}\right)^2}}. \tag{4.31}$$

**Fig. 4.59** Schematic representation of the projection process [49]

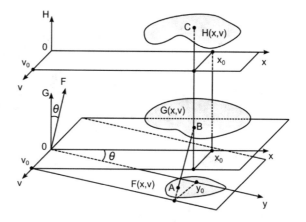

### 4.5.4   Application of Whole-Field Projection Moiré for the Registration of Radial Blood Flow Pulses

An artery goes underneath a skin surface, which is provided below (Fig. 4.60). To define this surface a function GK(x, v) is used:

$$G_K(x, v) = 0.1 \exp\left(-0.05\left(x^2 + v^2 - 2^2\right)^2\right) \tag{4.32}$$

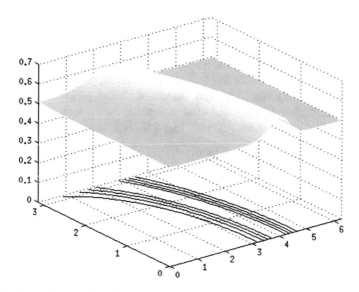

**Fig. 4.60** Illustration of surface described by function G(x)

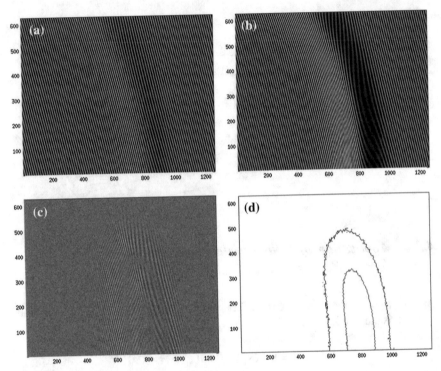

**Fig. 4.61** Images (*top-left*) and (*top-right*) shows projected grating on a surface G(x) and a deformed surface. Image (*bottom-left*) shows subtractive superposition of (*top-left*) and (*top-right*). Image (*bottom-left*) shows contour lines of a deformation

Illumination angle θ is equal to 450 at grating pitch λ of 0.05. The relief is initiated by a deformation, resulting from an impulse of pressure applied to the surface. Provide below (Fig. 4.61, top-left) is a grating projected on a provided surface G(x) and Fig. 4.61 top-right demonstrates the projected grating on a deformed the surface G(x). A subtractive superposition of the two surfaces is provided in Fig. 4.61 bottom-left with Fig. 4.61 bottom-right demonstrating the deformation contour lines.

## 4.6  Proposed Prototype of Wrist Watch-like Radial Pulse Analysis Sensor

### 4.6.1  Prototype Design

Executing experiments with the equipment thoroughly described in previous sections idea to create prototype of portable wrist watch-like sensor originated. It started with the understanding of the measuring process and capabilities of it in near future. The radial pulse analysis experiment roughly consisted of:

  Micro-membrane or micro-membrane matrix attached to the place of pulsation;

Laser beam;
Diffraction grating for laser light splitting;
Charge-coupled device (CCD) for registering reflected laser light;
Oscilloscope;
Computer for data representation and analysis.

  All the enumerated parts of equipment used in experiment were relatively small in size. Taking into account that micro sensors or nanotechnologies are becoming more important in everyday life and invoking statics it is obvious that significant miniaturizing of various types of sensors is just the question of time. Thus, it can be understood that all the devices used in the experiment today, in near future easily can be miniaturized and installed in single piece watch-like optical sensor. Such process will lead to portable, significantly comfortable, cost effective and easily usable radial pulse analysis sensor. Moreover, being wrist watch-like shape it could be used practically all day long, recording the information of pulse and storing it for further analysis, or in emergency case giving emergency signal.

#### 4.6.1.1   3D Computer Graphics Software

For prototyping Autodesk 3ds Max computer graphic software was used.

  Recognizing that such computer software has exclusive characteristics for the presentation of any idea or reality in professional manner it is clear that it is a must to master such software from engineering point of view.

### 4.6.2   Proposed Prototype Geometry

For the particular case the geometrical parameters of proposed radial pulse analysis sensor case is presented in Fig. 4.62.

  Referring to the experiment sensor essential parts is micro-membrane, CCD and laser. Figure 4.63 illustrates the shape, basic elements and application of sensor prototype.

  To get more sensitive signal membrane matrices can be introduced to the system (see Fig. 4.64).

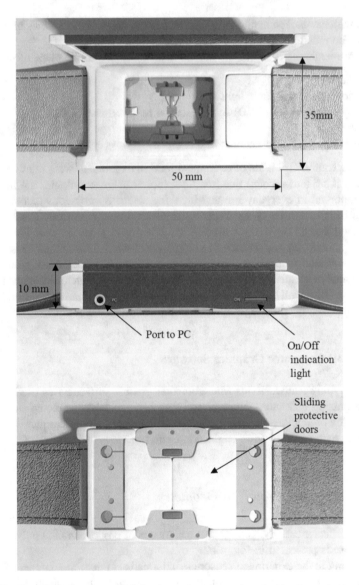

**Fig. 4.62** Geometry of prototype radial pulse sensor: **a** top view; **b** side view; **c** bottom view

Working principle of proposed sensor is based on laser triangulation described in previous chapters. The vital parts, i.e. laser, membrane matrices, CCD and diffraction grating could be obtained in relatively small size fitting the geometrical dimensions of prototype. Nevertheless, in order to fabricate such sensor basic obstacle to overcome is packaging of micro-membranes, since being vulnerable structures they may collapse when pressure is introduced. Moreover, signal

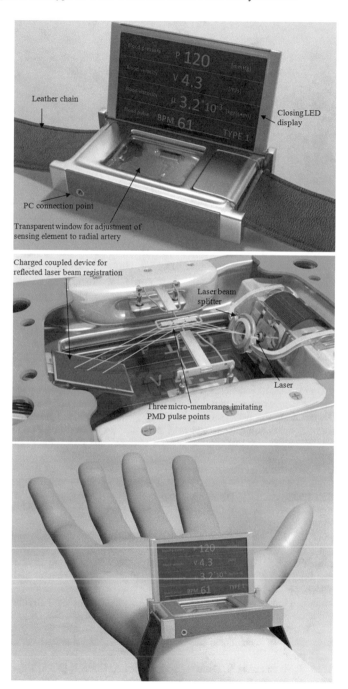

**Fig. 4.63** Prototype of radial pulse sensor: *top* basic elements (*top, side view*); *middle* basic elements (*undersurface view*); *bottom* sensor in action

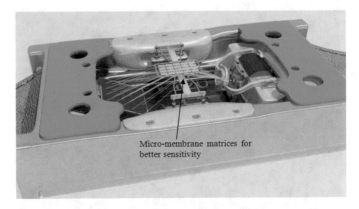

Micro-membrane matrices for
better sensitivity

**Fig. 4.64** Prototype of pulse sensor with micro-membrane matrices

calibration could be a difficult problem to solve. But as innovations are evolving it is reckoned that in near future such sensor could be introduced in everyday life giving real time monitoring of persons health registering blood pulse peculiarities.

# References

1. Sato T, Nishinaga M, Kawamoto A, Ozawa T, Takatsuji H (1993) Accuracy of a continuous blood pressure monitor based on arterial tonometry. Hypertension 866–886
2. Shenoy D, von Maltzahn WW, Buckley JC (1993) Noninvasive blood pressure measurement on the temporal artery using the auscultatory method. Ann Biomed Eng 21:351–360
3. Flaws B (1995) The secrets of Chinese pulse diagnosis. Blue Poppy Press, Boulder, USA, pp 1–49
4. Wang SH (1997) Thepulse classic. Blue Poppy Press, Boulder, pp 3–28
5. Lad DV (2005) Secrets of the pulse. The ancient art of ayurvedic pulse diagnosis. Motilal Banarsidass Publishers, New Delhi, pp 35–56
6. Upadhyaya S, Deva S, Nadi V (2005) Ancient pulse science. Narayana Publisher, Toronto, pp 15–68
7. Li Hammer (2001) Chinese pulse diagnosis. A contemporary approach. Eastland Press, Seattle, pp 105–128
8. Thakker B, Vyas LA (2010) Radial pulse analysis at deep pressure in abnormal health conditions. In: Third international conference on biomedical engineering and informatics, Yantai, China, pp 10–14
9. Malinauskas K, Ostasevicius V, Dauksevicius R, Grigaliūnas V (2012) Residual stress in a thin-film microoptoelectromechanical (MOEMS) membrane. Mechanika 18(3):273–279
10. Starman L, Busbee J et al (2001) Stress measurement in MEMS devices. In: Technical proceedings of the 4th international conference on modeling and simulation of microsystems, Hilton Head, pp 394–401
11. Ostasevicius V, Malinauskas K, Janusas G, Palevicius A, Cekas E (2016) Development and investigation of MOEMS type displacement-pressure sensor for biological information monitoring. In: Optical sensing and detection IV, Book Series: proceedings of SPIE, vol 9899
12. Judy JW (2010) Microelectromechanical systems (MEMS): fabrication, design and applications. Smart Mater Struct 10:1115–1134

13. Petersen KE (2005) Micromechanical membrane switches on silicon. IBMJ Res Develop 23:376–385
14. Madou JM (2002) Fundamentals of microfabrication: the science of miniaturization. CRC Press, USA, pp 131–469
15. Syms RRA, Moore DF (2002) Optical MEMS for telecommunications. Mater Today 26–35
16. Facilities Manual (1980) Reactive ions etch system PK-2420RIE. Plasma-Therm, Inc., Data sheet no. PK581
17. Reactive ion etching equipment. Website of Institute of Materials Science of Kaunas University of Technology. Access via internet: http://www.fei.lt/en/tech_6.htm
18. Elot E et al (2002) Metallic and other coating—Vickers and Knoop microhardness tests 35–42
19. Suslov A (2008) Atomic force microscope NT-206: operating manual. Microtestmachines Co, Gomel, p 5
20. Clemente-Vilalta A, Gloystein K (2008) Principles of atomic force microscopy (AFM). Aristotle University, Thessaloniki, pp 1–2
21. EasyScan (2005) 2 AFM operating instructions, version 1.3. Nanosurf AG, Switzerland, pp 8, 10, 12–17, 34
22. Hu JN, Yan SC, Wang XZ, Chu H (1997) Anintelligent traditional chinese medicine pulse analysis system model based on artificial neural network. J China Med Univ 26(2):134–137
23. Dharmananda S (2000) The significance of traditional pulse diagnosis in the modern practice of Chinese medicine. Institute for Traditional Medicine, Portland, Oregon. http://www.itmonline.org/arts/pulse.htm
24. Shu JJ, Sun Y (2006) Developing classification indices for Chinese pulse diagnosis. Complement Ther Med 15(3):190–198
25. Gore WL & Associates, Inc. AP2400-EN124GORE Hybrid Vascular Graft. http://www.goremedical.com/hybrid/technical/
26. Keyence America. http://www.keyence.com/products/measure/laser/lkg/lkg.php
27. MTI Instruments Inc. http://www.mtiinstruments.com/technology/triangulation.aspx
28. Malinauskas K, Ostasevicius V, Dauksevicius R, Jūrėnas V (2012) Non-invasive micro-opto-electro-mechanical system adaptation to radial blood flow pulse and velocity analysis. Mechanika 18(4):479–483
29. Smith HM (1975) Principles of holography, 2nd edn. Wiley, New York
30. Caulfield HJ (1979) Handbook of optical holography. Academic Press, Massachusetts
31. Hariharan P (1984) Optical holography. Cambridge University Press, Cambridge
32. Saxby G (1988) Practical holography. Prentice-Hall, New York
33. Niyogi P, Chakrabartty KS, Laha KM (2005) Introduction to computational fluid dynamics. Cambridge University Press, New York, pp 57–64
34. Bergel DH (1961) The static elastic properties of the arterial wall. J Physiol 156(3):445–457
35. Carew ET, Vaishnav NR, Patel JD (1968) Compressibility of the arterial wall. Circ Res 23:61–68
36. Agarwal R, Katiyar VK, Pradhan PA (2008) Mathematical modeling of pulsatile flow in carotid artery bifurcation. Int J Eng Sci 46(11):1147–1156
37. Figueroa AC, Baek S, Taylor AC, Humphrey DJA (2009) Computational framework for fluid-solid-growth modeling in cardiovascular simulations. Comput Methods Appl Mech Eng 198(45–46):3583–3602
38. Burton CA (1954) Relation of structure to function of the tissues of the walls of blood vessels. J Physiol 34(4):619–642
39. Chen HYH, Sheu HWT (2003) Finite-element simulation of incompressible fluid flow in an elastic vessel. Int J Numer Methods Fluids 42(2):131–146
40. Chuong JC, Fung CY (1984) Compressibility and constitutive equation of arterial wall in radial compression experiments. J Biomech 17(1):35–40
41. Flori F, Giudicelli B, Di Martino B (2009) A numerical method for a blood-artery interaction problem. Math Comput Model 49(11–12):2145–2151
42. Fortin A, Fortin M (1990) A preconditioned generalized minimal residual algorithm for the numerical solution of viscoelastic fluid flows. J Non-Newton Fluid Mech 36:277–288

43. Liepsch WD (1986) Flow in tubes and arteries—a comparison. Biorheology 23(5):395–433
44. Oscuii NH, Shadpour TM, Ghalichi F (2007) Flow characteristics in elastic arteries using a fluid-structure interaction model. Am J Appl Sci 4(8):516–524
45. Taylor AC, Hughes RJT, Zarins KC (1998) Finite element modeling of blood flow in arteries. Comput Methods Appl Mech Eng 158(1–2):155–196
46. Kobayashi AS and Society for Experimental Mechanics (U.S.) (1987) Handbook on experimental mechanics. Prentice-Hall, Englewood Cliffs
47. Breque C, Dupre JC, Bremand F (2004) Calibration of a system of projection moiré for relief measuring: biomechanical applications. Opt Lasers Eng 41(2):241–260
48. Lehmann M, Jacquot P, Facchini M (1999) Shape measurements on large surfaces by fringe projection. Exp Techn 23(2):31–35
49. Malinauskas K, Palevicius P, Ragulskis M, Ostasevicius V, Dauksevicius R (2013) Validation of noninvasive MOEMS-assisted measurement system based on CCD sensor for radial pulse analysis. SENSORS 13(4):5368–5380

# Chapter 5
# Microsystems for the Effective Technological Processes

**Abstract** Piezoelectric composite material whose basic element is PZT is developed. It allows controlling the optical parameters of the diffraction grating, which is imprinted in it. In addition, the sensitivity of the designed element is increased using silver nanoparticles, thus surface plasmon resonance effect appears. The developed technological route of the production of complex 3D microstructure, from designing it by the method of computer generated holography till its physical 3D patterning by exploiting the process of electron beam lithography and thermal replication, is presented. The experimental technology for the better quality of complex microstructure replicas based on high frequency excitation in the mechanical hot imprint process was proposed and implemented.

## 5.1 Periodical Microstructures Based on Novel Piezoelectric Material for Biomedical Applications

Past few decades biomedical applications have required fast, reliable, miniature and low-cost methods and tools for recognition of bio-molecules in various fluids. One of the recent new applications in this area is related to bio-sensing elements based on cantilever type sensing platforms. These platforms are able to convert biological responses into electrical signals [1, 2]. Great potential in bio-sensing application areas have been found not only with cantilever sensors [3–5] but also with film bulk acoustic resonators (FBAR) [6].

This chapter discusses a design of a cantilever-type sensing element made of a novel piezoelectric material exhibiting high resonance frequencies, leading to a faster response time and much higher sensitivity compared to cantilevers made of silicon [7], silicon oxide [8] and etc. The advantage of the proposed design in this paper is a periodical microstructure imprinted on top of the piezoelectric layer with metal nanoparticles precipitated on the grating ridges. Because of incorporation of noble (in this case silver) nanoparticles, the Surface Plasmon Resonance (SPR) effect appears. This effect highly influences the efficiency, the structure and operation itself, i.e. much greater control of optical properties, sensitivity and

© Springer International Publishing AG 2017
V. Ostasevicius et al., *Biomechanical Microsystems*, Lecture Notes
in Computational Vision and Biomechanics 24,
DOI 10.1007/978-3-319-54849-4_5

selectivity may be achieved. Moreover, to achieve the maximum optical effect, an operating wavelength of the sensing element may be tuned to a spectral region where SPR peak is sharpest. SPR is a powerful tool for investigating biomolecular interactions with label-free real-time analytical technologies.

Many various materials have a property of piezoelectricity, but only few of them are most promising in MEMS and NEMS technologies, i.e. zinc oxide (ZnO) films [9, 10], polyvinylidene fluoride (PVDF) films [11, 12] and lead zirconate titanate (PZT) [13, 14], so-called, polycrystalline ceramics. These three main materials have high piezoelectric coefficients, a very good flexibility and a strong electrome-chanical coupling. Moreover, it is known that nano-sized ceramics are very different compared to bulk ceramics in their mechanical behavior. This research paper was concentrated on designing a novel piezoelectric material, working at low fre-quencies and able to harvest energy or to cause deformations. This novel material may be integrated in sensing or actuating elements, depending on the purpose of microsystem. Economical and easy fabrication allows it to use in nowadays tech-nologies. In this research paper, PZT (exhibiting high piezoelectric coefficient and permittivity, large dielectric constants and good conversion efficiency) was chosen as a basic material for creating a novel piezoelectric sensing platform. A mixture of 20% polyvinyl butyral together with PZT powder was synthesized. Since property of strong binding is essential, a polyvinyl butyral works here as a binding element with PZT ensuring good adhesion and flexibility. Three concentrations of PZT (40, 60 and 80%) composite materials were produced for profound investigations. Each PZT material was coated as a thin film and sandwiched between two cooper electrode layers. Results showed that in the mode configuration of d31, it harvested energy, i.e. at 50 Hz frequency it generated up to 80 µV. Further, a 4 micron periodical microstructure was imprinted on thin 80% PZT piezoelectric film and silver (Ag) nanoparticles deposited on it. A property of piezoelectricity allows tuning a diffraction grating and results in variation of diffraction efficiencies. It allows achieving desirable spectral region in which the designed cantilever-type sensing platform would be of a high-efficiency. To perform certain specialized sensing functions it must reliably store and convert different forms of energy, transduce signals, and respond repeatedly to external biological and chemical environments. The designed cantilever-type sensing platform can alter mechanical stress within the oscillator and its total mass when target analyte is in contact with its surface. Here, a signal transduction is achieved by employing a diffraction grating to measure the mechanical bending or the frequency spectra resulting from additional loading by the absorbed mass. Since both the resonant frequency shift and deflection are highly dependent on the position of the absorbed material (an-alyte) it is difficult to determine the exact amount of additional mass. A diffraction grating and incorporation of Ag nanoparticles on its surface allow adequate control of chemical surface functionality for the detection of analytes of interest, i.e. defined molecules can be absorbed from analyte onto the Ag anchors, creating a strong interaction between the functional group and the silver nanoparticles. Vibrating cantilever-based platform offers quantitative assessment of the specific mass when experimentally monitoring resonant frequency shifts. Advantages of the designed novel sensing platform include: easy fabrication; inexpensive materials and equipment required; ability to make thin, from 400 nm to 1.4 μm, films; wide application areas (from sensitive diagnostic devices in medicine, pharmacy, to microsystem devices in wireless sensors and portable electronics). Interesting fact is that few different components—resonance, piezoelectricity, diffraction efficiency and silver nanoparticles, all are combined in a single element.

## 5.1.1 Synthesis and Formation of PZT Coating

An oxalic acid/water based for synthesis nano powders of lead zirconate titanate (Pb $(Zr_x, Ti_{1-x}) O_3$) with x = 0.52, also known as PZT (52/48), was used. The precursors of PZT (52/48) solution were lead(II) acetate [$Pb(NO_3)_2$], titanium butoxide [$Ti(C_4H_9O)_4$], and zirconium butoxide ($Zr(OC_4H_9)_4$). The other reagents used were oxalic acid, deionized water, acetic acid, and ammonia solution. Lead (II) acetate [$Pb(NO_3)_2$] (8.26 g) was poured into 100 ml of water. Then, acetic acid was added and the solution was heated to 50 °C and mixed to dissolve. In 500 ml of water, 32 g of oxalic acid was dissolved, then stirred with the titanium butoxide (5.1 g) and zirconium butoxide (7.65 g) at a concentration of 80%. Afterwards, the lead acetate solution was added to the titanium butoxide and zirconium butoxide solution. The final solution was alkalised with 25% ammonia solution till pH 9–10 and mixed for an hour. The precipitate of the solution was filtered in vacuum and during filtering it was washed with water and acetone. After filtering, the material was dried at 100 °C for 12 h. The powder was heated at 1000 °C for 9 h. Finally, PZT powder was milled and mixed with 20% solution of polyvinyl butyral in benzyl alcohol mixed under defined conditions. Three material types using different PZT (40, 60 and 80%) concentration were produced. Finally, the paste was coated on a copper foil using a screen—printing technique.

Three different types of polyester monofilament screen meshes were used: 32/70, 48/70, 140/34. The coating was then dried in the furnace for 30 min at 100 °C. Different size of a screen mesh was chosen for controlling the thickness of a PZT coating. Thus, three coatings of different thickness were formed and investigated: element 1 with PZT coating of 68 μm thickness, element 2 with 60 μm thickness and element 3 with 25 μm thickness (see Table 5.1).

**Table 5.1** Properties of screen mesh and layer thickness

| Meshed screen type | Mesh opening (μm) | Thread (μm) | Open area (%) | Mesh thickness (μm) | Theoretical ink volume (cm³/m²) | Formed PZT layer thickness (μm) |
|---|---|---|---|---|---|---|
| 32/70 | 245 | 70 | 60.5 | 108 | 65 | 68 ± 1 |
| 48/70 | 130 | 70 | 42.3 | 107 | 46 | 60 ± 1 |
| 140/34 | 30 | 34 | 22 | 52 | 11 | 25 ± 1 |

**Fig. 5.1**  Pole alignment set

Further, a four micron periodical microstructure was formed on the top of piezoelectric thin film by hot-embossing technique. Ag nanoparticles were formed from a solution of 0.05 M AgNO$_3$ in deionized water and dip-coated on the periodical microstructure.

Before measuring the generated voltage, an electrical pole alignment was applied on a PZT coating. It was accomplished with a high voltage generator and a custom-made holder, shown in Fig. 5.1. An element with a PZT coating was placed in the special holder between positive and negative poles. The high voltage generator was set at 5 kV current and held for 30 min. The poling technique aligns a positive pole on one side of the PZT coating and a negative pole on the other side. This process improves voltage characteristics of a piezoelectric coating.

A novel sensing element was made of a cantilever-type (Fig. 5.2).

A cantilever based sensing platform allows precise evaluation of piezoelectric properties. The elements were investigated in both—direct and indirect piezoelectric effects.

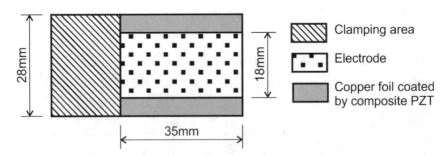

**Fig. 5.2**  Principal scheme of piezoelectric element

## 5.1.2 Characterization Methods

**Structural and Chemical Composition Measurements**

The structure and chemical composition of the designed material was investigated using Scanning Electron Microscope (SEM) Quanta 200 FEG. Which is also integrated with the Energy Dispersive X-Ray Spectrometer (EDS) detector X-Flash 4030 from Bruker. Samples were examined under the atmosphere of a water steam of controlled pressure. A 133 eV (at Mn K) energy resolution at 100.000 cps was achieved with a 30 mm$^2$ area solid state drift detector, cooled with Peltier element. The X-Ray spectroscopy method allows analyzing energy distributions. The energy differences were measured between various quantum states of the system together with the probabilities that the system jumps between these states.

FTIR—Fourier Transform Infrared Spectroscopy (a system SPECTRUM GX 2000 RAMAN) was used for the investigations of changes in chemical composition when the coating was poled and not poled. The diapason of FTIR spectrum was— 10,000–200 cm$^{-1}$. This technique enabled the researchers to identify changes in chemical compounds of the elements.

For qualitative and quantitative analysis of chemical compounds an X-ray diffractometer D8 Discover (Bruker) was used. The atomic and molecular structure of designed PZT was identified.

**Evaluation of Surface Morphology**

The investigations of surface morphology were performed with Atomic Force Microscope NT-206 in contact mode. Atomic force microscopy is a surface analytical technique used to generate very high-resolution topographic images of surface down to molecular/atomic resolution, the sample deposited on a flat surface being the only requirement. Depending on the sharpness of the tip it gives spatial resolutions of 1–20 nm. It can record topographic images as well as providing some information on nanoscale chemical, mechanical (modulus, stiffness, viscoelastic, frictional), electrical and magnetic properties when using specialized modes. Morphology parameters are as follows: $Z_{mean}$—average height, $R_a$—arithmetic average surface roughness, $R_q$—root mean squared surface roughness.

## 5.1.3 Dynamic Investigations of PZT Coatings

**Harmonic Excitation Measurements**

Harmonic excitation measurements were performed using a scheme, depicted in Fig. 5.3. It consists of a piezoelectric element, excitation measurement systems and data acquisition. An electromagnetic shaker excites the element fixed in a custom-made clamp. A harmonic excitation signal is transmitted to the electromagnetic shaker and controlled by a function generator AGILENT 33220 A and a voltage amplifier HQ Power VPA2100MN. For acceleration measurements, at the top of the clamp a single-axis miniature accelerometer KD-35 (with sensitivity of

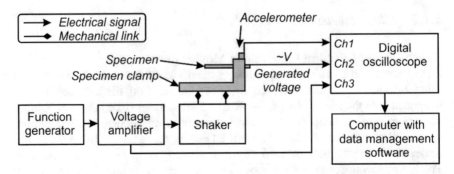

**Fig. 5.3** An experimental setup scheme used to measure harmonic excitations of piezoelectric elements

50 mV/g ±2% and working frequency from 5 Hz to 5 kHz) is attached. Signals from a voltage amplifier, accelerometer and an element are collected with data acquisition system, consisting of a 4-channel USB oscilloscope (analog-to-digital converter) PICO 3424, and forwarded to a computer. Obtained data is then analyzed with Pico-Scope 6 software.

**Impulse Excitation Measurements**

The experimental setup of investigations consists of a piezoelectric energy harvester (PVEH) applying single hits, excitation, and measurement systems and data acquisition. Data acquisition system consists of a 4-channel USB oscilloscope (analogous-to-digital converter) PicoScope 6000 series (Fig. 5.4) that collects signals from the accelerometer and PVEH. Signals from the oscilloscope are forwarded to the computer and managed with software PicoScope 6000. The system is based on a mathematical pendulum (Fig. 5.4).

The experimental system was designed as a mathematical pendulum which provides a single impulse to the clamped element when indicated. Sensor—head LK-G82 was used (with its accuracy 0.2 mkm). The response of vibrations was sensed with a laser triangular displacement sensor LK-G3000 (Keyence, Illinois, USA) and the measured data was collected with data acquisition system PicoScope (with data reading velocity 5 Gs/s).

**Electrical Excitation Measurements**

A two beam speckle pattern interferometer, or so-called PRISM system (measurement sensitivity <20 nm, measurement range >100 μm), was used to evaluate a response of electrical excitation of the investigated piezoelectric element. This method allows measuring vibration and deformation with minimal sample preparation and with no contact with the sample surface. PRISM is a rather high-speed holographic technique, equipped with a computer system and integrated software (Fig. 5.5).

**Fig. 5.4** Experimental setup consisting of: (*1*) a mathematical pendulum, (*2*) a sensor head LK–G82, (*3*) an investigated element, (*4*) PicoScope oscilloscope, (*5*) control block LK–G3001PV and (*6*) a power supply block

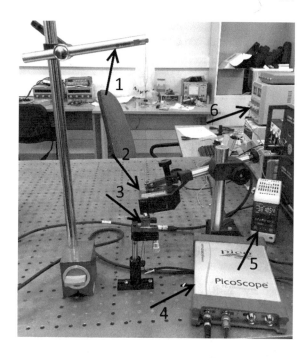

**Fig. 5.5** A PRISM measurement system consisting of: (*1*) a control block, (*2*) an object illumination head, (*3*) a video camera

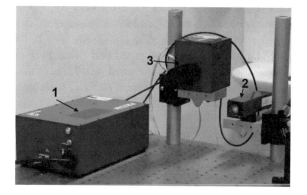

The laser beam directed to the object is an object beam; the beam directed to the camera—a reference beam (Fig. 5.6). Camera lens collects the scattered laser light from the object and images the object onto sensors of CCD camera. The reference beam falls directly to the camera and overlaps the image of the object. The fringes displayed on the monitor appear because of the shape changes occurring between a reference and a stressed state of the object. Obtained data allows evaluating the electrical excitations of the investigated element.

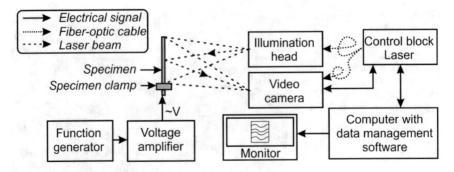

**Fig. 5.6** The PRISM experimental set-up

### 5.1.4  Structure and Chemical Composition of PZT Composite Material

Dispersive X-Ray Spectrometer was used here for energy distribution analysis.

Figure 5.7 shows the XRD pattern of PZT powder after the final calcinations process. PZT ceramics crystallises in a tetragonal structure (a = b = 4.006 Å, c = 4.128 Å, $\alpha = \beta = \gamma = 90°$) with space group P 4 mm (noncentrosymmetric) and the (0 0 1), (1 0 0), (1 0 1), (1 1 0), (1 1 1), (0 0 2), (2 0 0), (1 0 2), (2 1 0), (1 1 2), (2 1 1), (2 0 2), (2 2 0), (1 0 3), and (3 2 0) crystallographic plane orientations which correspond to values reported in [14, 15]. XRD pattern of PZT powder corresponds to $Pb(Zr_{0.52}Ti_{0.48})O_3$ with R–f of 0.31.

FTIR analysis of a non-poled and poled PZT coating applied using different screen-printing meshes was carried out. There is no significant influence of poling and thickness upon the spectrum of PZT; therefore, the typical FTIR absorbance spectrum at 4000–500 $cm^{-1}$ of PZT coating is presented in Fig. 5.8.

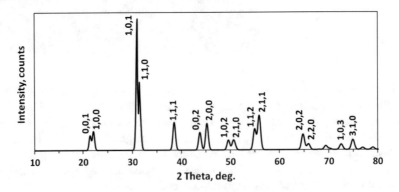

**Fig. 5.7** XRD pattern of PZT powder after the final calcinations process

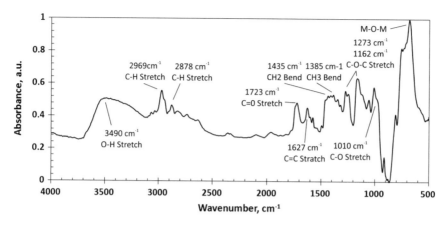

**Fig. 5.8** FTIR absorbance spectra with functional groups of PZT coating

In the FTIR spectra, the strong and broad absorption peaks were observed at 3490 cm$^{-1}$ (O–H stretch), 2969 cm$^{-1}$ (C–H stretch), 2878 cm$^{-1}$ (C–H stretch), 1723 cm$^{-1}$ (C=O stretch), 1627 cm$^{-1}$ (C=C stretch), 1435 cm$^{-1}$ (CH$_2$ bend), 1385 cm$^{-1}$ (CH$_3$ bend), 1273 cm$^{-1}$ (C–O–C stretch), 1162 cm$^{-1}$ (C–O–C stretch) and 1010 cm$^{-1}$ (C–O stretch). The entire array of these peaks corresponds to the FTIR absorbance spectra of PVB [15]. A wide and strong peak observed in the range of 800–550 cm$^{-1}$ corresponds to the M–O–M bonds (M is metal) of PZT, e.g. Ti–O, Ti–O–Ti, Zr–O and Zr–O–Zr [16].

Chemical compositions of PZT composite materials of different concentration were investigated with energy-dispersive (ED) spectrometer. A pulse height analysis is employed. Ionization is caused by incident X-ray photons, electrical charge is produced. Energy dispersive spectrum is displayed. The x-axis represents the X-ray energy in channels 1.5–5 eV wide and the y-axis represents the number of counts per channel up to 1600 cps/eV. Three main elements were defined in the samples: Lead (Pb), Zirconium (Zr), and Titanium (Ti). Conventionally, for the Ti Kβ energy resolution peak is specified at about 4.94 keV. For Zr Lα the peak is achieved at ~2.05 eV and Pb peak is about 2.35 eV (Fig. 5.9).

The peak values of Pb, Zr and Ti of the elements (when PZT 40, 60 and 80%) for more accurate comparison are presented in Table 5.2.

Each element has its defined positions of characteristic peak. This peak corresponds to the transitions in its electron shell. Chemical analysis shows that Pb, Zr and Ti concentrations increases proportionally to concentration of PZT. Both Zr and Ti show a dominant K and L peaks, respectively. The Pb spectrum is more complex. Its dominant peak is observed at ~2.35 keV. The intensity value is dependent on the exciting X-rays intensity, on the energy spectrum, X-ray detector's efficiency and on the geometry of the investigated element and the source.

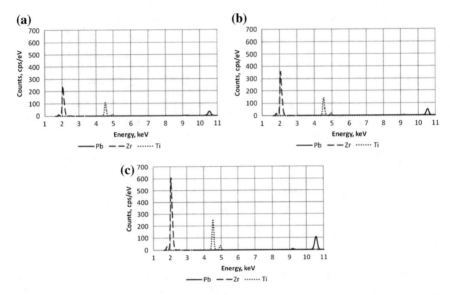

**Fig. 5.9** ED spectrum showing peaks of Pb, Zr and Ti of piezoelectric elements with PZT: **a** 40%; **b** 60% and **c** 80%

**Table 5.2** Peak values of Pb, Zr and Ti of designed piezoelectric elements

| Concentration (%) | Pb 10.5515 keV | Zr 2.04236 keV | Ti 4.50486 keV |
|---|---|---|---|
| 40 | 34.49 | 244.44 | 107.73 |
| 60 | 48.62 | 359.21 | 143.94 |
| 80 | 108.05 | 609.85 | 256.81 |

## 5.1.5   Surface Morphology of Novel Cantilever Type Piezoelectric Elements

Surface morphology of three different elements with PZT concentrations of 40, 60 and 80%, were investigated using Scanning Electron Microscopy (Fig. 5.10). Different sizes of grains are observed in samples: surface of the first element PZT 40% was rather smooth with small 5–15 µm diameter pileups observed on top (Fig. 5.10a). Increased PZT concentration to 60% leads to formation of individual grains (Fig. 5.10b). The surface of PZT 80% piezoelectric film became granular with smaller grain size below 4 µm (Fig. 5.10b). PZT particle loaded in a polyvinyl butyral might be the origin of the irregular shape, nucleation and growth in the solution, thus forming the smaller grains group. PZT is land structures (Fig. 5.10c) were formed where the granular grains surround larger grains. It is also seen that a high density is achieved in PZT 40 and 60% except few pinholes. However, these micro cracks with the length of micron distribute uniformly in the surface (Fig. 5.11).

(a)                    (b)                    (c)

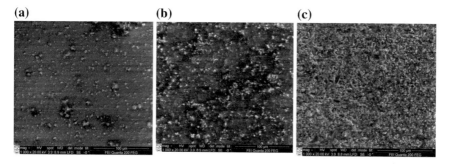

**Fig. 5.10** SEM view of piezoelectric elements with PZT concentration of **a** 40%, **b** 60% and **c** 80%

(a)                    (b)                    (c)

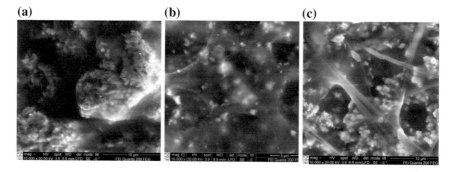

**Fig. 5.11** Images of SEM of piezoelectric elements when PZT: **a** 40%; **b** 60%; **c** 80%

SEM surface views in scale of 5–10 μm are presented in Fig. 5.11. The element with PZT concentration of 40% (Fig. 5.11a) has some small structures of about 10–12 nm sizes. The element with PZT 60% concentration has a small net with cavities formed on the surface (Fig. 5.11b). Higher PZT concentration leads to formation of three dimensional microstructures with empty cavities of about 6–8 μm diameter (Fig. 5.11c).

SEM views of the different thickness elements are given in Fig. 5.12. Results show that element 1 has small granular grains on the surface of a diameter ∼1.1 μm. Element 2 has a smoother surface with less grains of a diameter ∼0.9 μm. Element 3 has three dimensional structures with empty cavities of a 6–8 μm diameter.

Full composition of the elements was defined in Fig. 5.13. The main elements in the composition are Carbon (C) and Zirconium (Zr);—both are very good conductors defining good piezoelectric properties of designed novel coatings.

Atomic Force Microscopy evaluated the surface morphology of the designed elements (Fig. 5.14, Table 5.3). 3D views show that element 1 has a rather smooth surface with roughness $R_q = 29$ μm (Fig. 5.14a). Elements 2 and 3 have rough surfaces with roughness $R_q = 189$ μm and $R_q = 149$ μm, respectively (Fig. 5.14b, c).

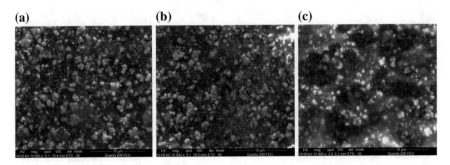

**Fig. 5.12**  SEM views of samples: **a** element 1 (mesh 32/70); **b** element 2 (mesh 48/70); **c** element 3 (mesh 140/34)

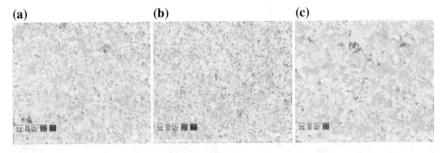

**Fig. 5.13**  Elemental mapping done with SEM of the **a** element 1 (mesh 32/70); **b** element 2 (mesh 48/70); **c** element 3 (mesh 140/34)

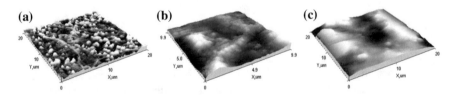

**Fig. 5.14**  AFM 3D view: **a** 32–70; **b** 48–70; **c** 140–34

Using a different screen-printing mesh allows controlling not only thickness but also surface morphology of the element.

## 5.1.6  Piezoelectric Properties

Piezoelectric elements' response to harmonic excitation was investigated using system presented in Sect. 5.1.3. Results showed, when exciting periodically (acceleration 0.007g, frequency 50 Hz) the designed cantilever type piezoelectric elements, they generate up to 50 µV potential (for open circuit) when PZT 40%

**Table 5.3** AFM values of the surface morphology

| Element | Mesh | $Z_{mean}$ | $R_a$ | $R_q$ |
|---|---|---|---|---|
| 1 | 32/70 | 54 | $21 \pm 1$ | $29 \pm 1$ |
| 2 | 48/70 | 396 | $156 \pm 0.5$ | $189 \pm 0.5$ |
| 3 | 140/34 | 457 | $112 \pm 0.5$ | $149 \pm 0.5$ |

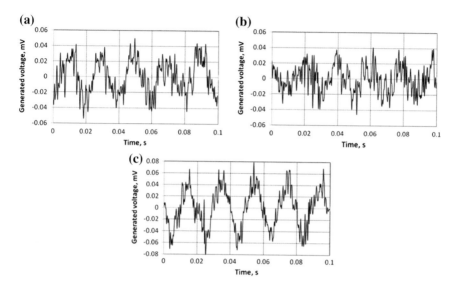

**Fig. 5.15** Results of electric potential generated by designed elements with PZT: **a** 40%; **b** 60% and **c** 80%

(Fig. 5.15a), and up to 40 µV potential when PZT concentration is 60% (Fig. 5.15b). The element with 80% concentration of PZT generates electrical potential up to 80 µV. Results were preprocessed by low-pass filter of 500 Hz.

Cantilever-type piezoelectric element with 80% concentration of PZT shows significant results in power generation as a thin layer at low frequencies. Other elements had no signs of ability to convert electrical potential into mechanical energy. The element with PZT 80% was investigated using interferometer PRISM of electronic speckle pattern (Fig. 5.16).

The designed element (PZT 80%) was excited by a sinusoidal function with an amplitude of 5 mV at a frequency of 13 Hz (Fig. 5.15). At the first resonant frequency the element vibrates as a clamped-free cantilever resonator in its fundamental flexural mode.

Significant advantage of this element is an ability to apply designed novel piezoelectric composite material at any thickness, form and size on any uniform or non-uniform vibrating surface.

Pole effect on electrical properties of PZT composite material were investigated with PVEH based on direct piezoelectric effects. However, the investigations based

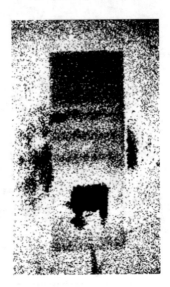

**Fig. 5.16** Hologram of the cantilever-type piezoelectric element with PZT 80%

on the direct piezoelectric effect showed remarkable results, i.e. under impulse force of 5 N amplitude applied on the poled element, it generated from ∼1.4 to ∼3.5 mV (Fig. 5.17). Poled element 3, the one with the thinnest PZT layer, generated the highest voltage of 3.6 mV (Fig. 5.17c). It was easy to detect the difference between poled and not poled elements, i.e. around 61% of a less generated voltage.

During poling the material is subjected to a very high electric field that orients all the dipoles in the direction of the field. Upon switching off the electric field most dipoles do not return back to their original orientation as a result of the pinning effect, produced by microscopic defects in the crystalline lattice. This gives material comprising numerous microscopic dipoles that are roughly oriented in the same direction.

The novelty is in the designed material PZT, i.e. obtained material is not classic PZT with wider application areas. The designed microresonator may be operated in

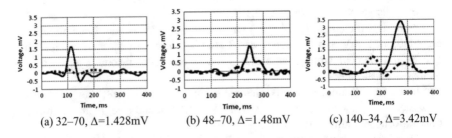

(a) 32–70, Δ=1.428mV        (b) 48–70, Δ=1.48mV        (c) 140–34, Δ=3.42mV

**Fig. 5.17** Generated voltage diagram of **a** element 1 (*straight line* poled, *black square* not poled); **b** element 2 (*straight line* poled, *black square* not poled); **c** element 3 (*straight line* poled, *black square* not poled)

a system with an indicated resonant frequency by varying dimensions of the microresonator's layers and its geometrical parameters. The aim of this research was to create a novel microresonator with controllable parameters which could assure much higher functionality of MEMS. Creation of this novel element will allow it to integrate in various MEMS systems: high stability electric oscillation sources (as generators), electric filters, in energy harvesting, sensors for testing proteins, viruses, chemical species and etc.

### 5.1.7 Calculation of Module of Elasticity

Using PicoScope oscilloscope the experimental vibrational response curve has been registered. In order to compare the experimental and theoretical models, the natural frequency of proposed multilayer specimen had to be calculated. To calculate the natural frequency formulae (5.1), (5.2), and (5.3) had to be taken into account.

$$\omega_d = \omega_n \sqrt{1 - \xi^2} \tag{5.1}$$

where $\omega_d$—the damped natural frequency, $\omega_n$—the natural frequency, $\xi$—the damping coefficient.

$$\xi = \frac{1}{\sqrt{1 + \left(\frac{2\pi}{\delta}\right)^2}} \tag{5.2}$$

where $\xi$—the damping ratio, $\delta$—the logarithmic decrement.

$$\delta = \frac{1}{n} \ln \frac{x(t)}{x(t + nT)} \tag{5.3}$$

where $\delta$—the logarithmic decrement, $n$—number of subsequent periods, $x(t)$—amplitude at time $t$, $x(t + nT)$—amplitude at time $t + nT$, $T$—period.

The values of amplitudes were taken from the vibrational response curve of proposed multilayer structure (Fig. 5.18).

From Fig. 5.18 the damped natural frequency of 186 Hz was calculated. Measured vibration amplitudes in the measuring interval were 150 µm and after one period 60 µm. Using formulas stated in previous section the damping coefficient and natural frequency of proposed multilayer element was estimated. The natural frequency of the system was 187.056 Hz.

Experimental results were compared with numerical calculations (Fig. 5.19). Model of multilayer element gave similar results. Figure 5.19 shows the first mode of vibrations of the modelled structure.

The first form of vibrations in COMSOL Multiphysics model was achieved at natural frequency of 187.867 Hz. It was only 0.43% difference between theoretical

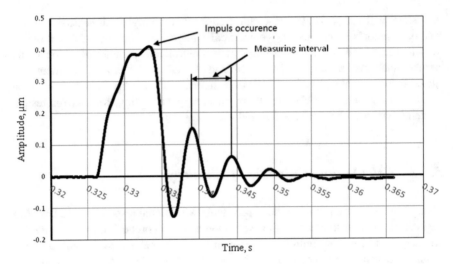

**Fig. 5.18** Vibrational response of multilayer element into single impulse

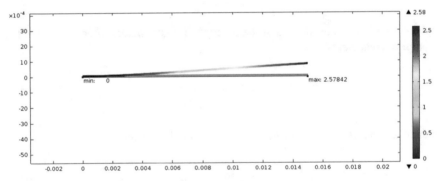

**Fig. 5.19** First vibration form of multilayer element model in COMSOL Multiphysics environment

and experimental values. As the natural frequencies of both models were equal, it could be stated that Young's modulus of created piezoelectric material of 0.66 MPa was determined correctly.

Due to the small Young's modulus, it is possible to create microresonators, which are able to operate in low frequency range allowing to increase their sensitivity limit11, while high resonant frequencies have several disadvantages as overshooting and settling time12. Comparing regular PZT material and proposed composite material the first resonant frequency decreases 2.3 times (from 439.387 to 187.867 Hz).

Experimental results also show that the created multilayer element can be used for energy harvesting. Multilayer structure oscillated by single mechanical impulse deforms the piezoelectric layer, which converts the kinetic energy into electrical energy. Piezoelectric layer is able to generate up to 12 mV voltage. Later dependence of generated voltage on different substrates, thickness of piezoelectric layer, and its polarization and binder polymer will be investigated.

## 5.1.8   Periodical Microstructure and SPR

A novel cantilever-type piezoelectric element (PZT 80%) works in both, direct and indirect, piezoelectric regimes. The designed novel piezoelectric composite is a promising material for future experimentations. This unique property allows a real time and direct observations of affinity interactions, i.e. sensing elements with piezoelectric effect employs the active method for measurements in medical or pharmaceutical fields. Imprint of periodical microstructure enables to use this platform for studying bio-molecular interactions, to analyze functional information, i.e. the information related to physiological effect of an analyte on a living system [15, 16]. It is essential in many important applications: medicine, pharmacy, cell biology, environmental measurements, etc. For this purpose, a four micron periodical microstructure was imprinted on formed thin film, designed from 80% concentration of PZT (element, exhibiting best piezoelectric properties). Schematic view of the designed element is given in Fig. 5.20. The platform consists of a thin piezoelectric film coated on a copper foil working as an electrode. Opposite electrode is formed on the top of a thin film. Periodical microstructure was formed by hot-embossing procedure at defined conditions.

Surface morphology of imprinted grating was measured using Atomic Force Microscopy. As previous researches showed, it is rather hard to imprint periodical microstructure in piezoelectric layer because of its brittleness and inelasticity. Here, an additive polyvinyl butyral was chosen to improve these properties. Thus, a well-defined grating was formed (Fig. 5.21). Average grating depth was ∼580 nm with an average period of 3.8 μm and rather smooth surface–surface roughness $R_q = 129.8$ nm.

Main drawbacks of such a sensing platform arise because of the interaction of biomaterials in the investigated analyte and are strongly influenced by the

**Fig. 5.20** A principal scheme of a cantilever-type piezoelectric sensing platform

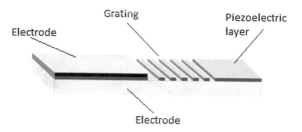

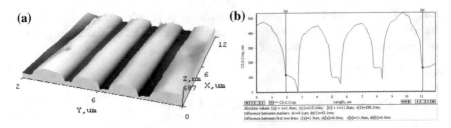

**Fig. 5.21** AFM view of a four micron periodical microstructure imprinted on a novel piezoelectric element **a** 3D grating view; **b** topography cross section of the grating

adsorption of bio-molecules, their diameter size and viscosity of the analyte. To overcome these drawbacks, silver nanoparticles were precipitated on the grating. The SPR enhances the absorption (optical signal); and nanoparticles are used as biological tags for quantitative detection of bio-molecules in analyte. Moreover, a combination of piezoelectric and SPR properties in a single element is an effective way to expand the working range of the element. To prove the relevance of Ag nanoparticles in the designed cantilever-type sensing platform diffraction efficiency measurements were taken (Fig. 5.22a). AFM surface view of the grating (Fig. 5.22b) showed that approximate size of Ag was $\sim 17$ nm diameter.

Diffraction efficiency measurements were performed using a laser diffractometer (Fig. 5.22a). A He–Ne red laser light was incident to the grating imprinted on a

**Fig. 5.22** *Top* diffraction efficiency measurement results, *bottom* AFM view of a grating top coated with Ag nanoparticles

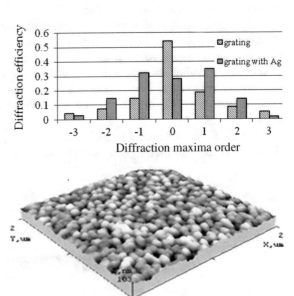

designed element (PZT 80%). Most of the diffraction energy was concentrated on its zero order ($\sim 54\%$) and in its first orders ($\sim 33\%$). For a periodical microstructure with silver nanoparticles, diffracted energy in its zero order has dramatically decreased to $\sim 28\%$ and in its first orders of its maximum increased up to 35%. In second orders of maximum it increased up to 29%. Thus, results imply that silver nanoparticles significantly improve the optical response of a novel sensing platform.

A future perspective of the designed novel cantilever-type piezoelectric element is related to its integration in MEMS for analysis of functional information as physiological effects of an analyte, type and concentration of molecules, and etc. When, for example, a constant potential is applied, electrochemical reactions may be observed together with the changes of mass or resonant frequency shifts. These measurements are often desired in biomedicine, cell biology, and environmental or pharmacy measurements.

Another very important advantage of the developed microresonator is the possibility to combine electrical and optical measurement techniques to increase accuracy of the sensing microresonator. Lamellar periodical microstructure (period —4 μm and depth—574 nm) was formed in PZT nanocomposite material using hot embossing technique at 100 °C temperature, 5 atm pressure and 10 s of holding time.

PZT nanocomposite material, used for formation of periodical microstructure, allows controlling not only mechanical properties of microresonator, but optical parameters too. Example of reflected diffraction efficiency of first order diffraction maxima and its diffraction angle dependence on period changes of microstructure caused by electrical signal is presented in Fig. 5.23. During this analysis the incident laser light (wavelength 632.8 nm) were cast at 40° angle to the top surface of nickel coated periodical microstructure integrated into PZT nanocomposite. Results have shown that the period changes by 10 nm leads to changes of diffraction efficiency to about 0.1% and diffraction angle about 0.15°.

Created microresonator with controllable mechanical and optical parameters assures much higher functionality of microelectromechanical systems.

**Fig. 5.23** Diffraction efficiency and diffraction angle versus period of periodical microstructure

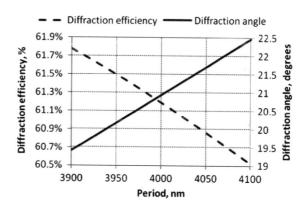

## 5.2   Development of Complex 3D Microstructures Based on Computer Generated Hologram

Health care monitoring is the most common application of sensors. Decade after decade they are becoming much smarter, more sensitive and have wider capabilities. Thus, today many sensing and actuating devices in micro scale are based on diffractive optics. Since light can be controlled by applying relatively low energy, the microstructures have been the essential component of the design of microdevices. Researches concentrated on the development of new manufacturing techniques for the fabrication of high quality periodical microstructures fulfilling the challenging demands of optical sensors [17, 18], biosensors [19, 20], synchrotrons, space missions, laser pulse compressors and other spectroscopic applications. Computer Generated Holography (CGH) is the digital method which is used to generate 3D microstructure. You can record hologram without using a real object when using CGH which is different from optical holography. The most popular techniques to calculate CGH are Gerchberg-Saxton algorithm [21] or Adaptive-Additive algorithm [22]. Both techniques employ classical iterative Fourier transform algorithm. CGH has been used for various applications, such as 3D displays [23], optical filters [24], optical testing [25], various encryption schemes [26, 27]. It has been also used for the development of wavefront sensor [28, 29]. In this section, computer generated holography (CGH) is employed for the creation of MOEMS element. Gerchberg-Saxton algorithm will be used to retrieve a phase data. There are a few techniques for the fabrication of periodical microstructures including mechanical ruling of gratings, holographic recording, and replication [30, 31]. The main aim of this section is to develop a technological route of the production of complex 3D microstructure, from designing it by the method of computer generated holography till its physical 3D patterning by exploiting the process of electron beam lithography and thermal replication.

Replication technology plays an important role in the production of microstructures. The most popular replication technologies are hot embossing, UV embossing, injection molding, and etc. They enable low-cost mass production of optical microstructures and nanostructures. Replication technologies reach extremely high resolution, while it could be merged with macro objects (millimeter or centimeter range). Therefore this technology could be applied for production of optical elements and modules. But good replication result depends not only from good quality of high-resolution mould [32]. Quality covers all steps of microstructures creation from the design to mass production, which are presented in Fig. 5.24.

The process starts with the design of the surface microstructure using optical design programs. The structure elements can vary from diffractive grating nanostructures to refractive microlenses and complex combinations of various elements. One of the more realization—computer generated hologram (CGH).

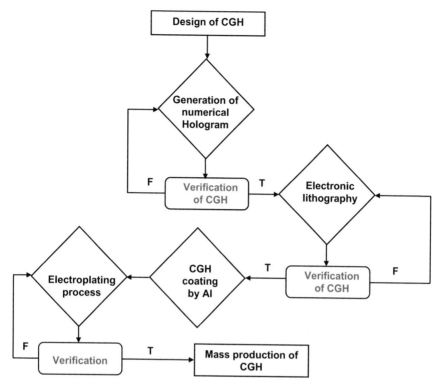

**Fig. 5.24** Algorithm for manufacturing of microstructure: from design of CGH to mass production

In the origination phase, the designed surface relief is fabricated using one of many technologies: laser writing [33, 34], e-beam writing [35], e-beam lithography [36] and holographic lithography [37] for feature resolution down to under 100 nm, to conventional lithography and etching for micrometer resolution.

The next step is the fabrication of a mold. The traditional silicon-based molds are brittle and have a limited longevity [38]. Metal, such as nickel stamper, has been widely used as micro/nano-scale molds [39]. The electroplating with nickel is widely used for making a metal copy from the microstructure in the original. Nickel is the most commonly used material for electroplating because it is significantly harder than, for example, copper or gold. The original structure must first be coated with a thin metal film to form a conductive coating for the subsequent electroplating. The metal film has a typical thickness of 10–100 nm, and typically, silver, gold, nickel or copper is used [40]. To maintain the quality of plated nickel, pH and the chemical composition of the electroforming bath must be continuously controlled. After the process, the master is separated from the original, and any residues are stripped with a suitable solvent.

The obtained nickel mold could be used as a master mold for the further replications of the microstructures using one of the main replication technologies:

- Injection molding
- Injection compression molding
- Hot embossing.

Criteria of these different molding techniques could be divided into several groups: the molding material used, the process technology used for replication and finally the cost effectiveness.

Based on the Gabor's holographic principles, nowadays a lot of optical phenomena and techniques are developed. Modern holography includes (a) the design of diffractive optical elements (DOEs) as a special case of CGHs applied for beam shaping, splitting and steering [41], optical interconnections [42], optical tweezers [43], multiphoton spectroscopy [44], lithographic fabrication of photonic crystals [45], and (b) digital holography (DH) like a useful method for non-destructive testing in: object contouring [46], biological microscopy [47], measurement of moving object [48], characterization of micro optical elements, system with CD and DVD [49], in X-ray microscopy [50].

Computer-generated holography (CGH), in contrast, calculates in the computer the reflection of light by a virtual object in the computer, the propagation of this reflected light to the hologram, and the interference with the reference light, to form a hologram.

The Fourier transformation (FT) is the mathematical tool and the foundation of the CGH. Most of the CGH mathematics is in two dimensions. The Fourier decomposition is useful in optics because exponential functions behave in a simple, well understood way as does re-composition of the functions after they have been acted upon by a optical system.

$$U(\xi, \eta) = \int_{-\infty}^{\infty} \int_{-\infty}^{\infty} u(x, y) e^{-2\pi i(\xi x + \eta y)} dx dy \qquad (5.4)$$

The correspondence of punctuality and periodicity can be extended to periodic objects which consist of regular arrays of points. The continuous Fourier transform of such objects will itself be periodic and consist of points. The relation between the point strengths of one period in the object and one period in the Fourier transform is given by the discrete Fourier transform (DFT):

$$U_{j^k} = \frac{1}{\sqrt{M}} \frac{1}{\sqrt{N}} \sum_{m=-\frac{M}{2}}^{\frac{M}{2}-1} \sum_{n=-\frac{N}{2}}^{\frac{N}{2}-1} u_{mn} e^{-2\pi i\left(\frac{jm}{M} + \frac{kn}{N}\right)}. \qquad (5.5)$$

There are several variations on defining the DFT. One variation is the normalization factors another in the range covered by the indices. This index variability is of practical importance.

The Fast Fourier transform (FFT) is actually an algorithm for performing the discrete Fourier transform or DFT. The algorithm speeds computation of the DFT by a factor $n/\log_2 n$ that can be considerable for large transforms [51].

In CGH, the method of calculating the interference fringes and the definition of the object shape are connected and cannot be considered independently. For calculation of the hologram, methods based on the Fast Fourier Transform (FFT), Fresnel integral, the point filling method and other are used in general. Of these, the method using the FFT can perform fast calculations but can be applied in principle only to planar objects. Yasuda et al. [52], Calixto et al. [33] simulated image obtained by calculating wavelength and intensity of diffracted light traveling toward the viewing point from the CGH. Wavelength and intensity of the diffracted light were calculated using FFT image generated from interference fringe data. Freese et al. [35] calculated the CGH phase function by an iterative Fourier transform algorithm (IFTA) [53]. Sakamoto and Nagao [54] used the patch model for CGH.

Many researches created a Lohmann-type CGH [55, 56]. In a Lohmann-type CGH based on the detour phase principle, the complex-amplitude data of a hologram are converted into amplitude and phase information. Amplitude is encoded by the area of a transparent aperture within each cell, whereas the phase is encoded by the position of the aperture. The complex amplitude $F(\mu, \upsilon)$ of a hologram is given by

$$F(\mu, \upsilon) = |H(\mu, \upsilon)|e^{j\phi(\mu,\upsilon)}, \tag{5.6}$$

where $H(\mu, \upsilon)$ and $\phi(\mu, \upsilon)$ are the amplitude and phase of a Fourier-transformed hologram, j is an imaginary unit [55].

Oikawa et al. [57] propose a fast CGH calculation method using an approximate Fresnel integral. In general, calculating Fresnel integrals requires a numerical integral; however, such a numerical integral consumes computational time. Therefore, when calculating a CGH using a Fresnel integral, it is difficult to calculate it in real-time.

Petruskevicius et al. [58] were obtained CGH using hologram simulation method based on the sources of point-spherical waves for objective wave and plane reference wave. The objective and reference wave field of the hologram plane $z = 0$ can be obtained as a sum of these sources. This can be expressed as follows:

$$E(x, y) = |E(x, y)|e^{i\phi(x,y)}, \tag{5.7}$$

where $|E(x, y)|$ is the modulus of the full field of complex electrical field strength and $\phi(x, y)$ is the full field phase distribution in hologram interfering plane. The distribution of the full field phase can be converted to a hologram grating relief, which is defined by the following generalized equation:

$$h(x, y) = H_{\max}\phi(x, y)/2\pi, \tag{5.8}$$

where $h(x, y)$ is the height of a relief grating point with coordinates x, y, which represents the grey level of CGH profile pixel and $H_{\max}$ is the maximum height of the relief grating. The maximum height of the grating pixel depends on the incident light wavelength ($\lambda$), pixel material refractive index (n):

$$H_{max} = \frac{\lambda}{n - 1},\tag{5.9}$$

corresponding to $2\pi$ phase shift of the transmitted wave. Resulting phase of the complete field $\phi(x, y)$ is reduced to a selected amount of grey levels in the $2\pi$ phase range to enable creation of the corresponding isosurfaces.

The CGH design techniques fall into two general categories: input/output techniques and iterative design techniques. All iterative design techniques are based on the phase retrieval algorithm proposed by Gerchberg and Saxton [53].

The schema of algorithm is shown in Fig. 5.25. These techniques are computationally efficient due to the use of a fast Fourier transform (FFT) [59]. For the original Gerchberg-Saxton algorithm, the phase constraint is that the element is phase only, the amplitude is normalized to one at each iteration, and directly quantized to the number of phase levels required in the design. The output constraint forces the target to match the desired output intensity leaving the phase of the output unchanged. The major problems with the Gerchberg-Saxton algorithm for diffractive optic design are that it is dependent on the initial phase estimate and stagnates readily due to the direct quantization of the phase profile [60].

Thomson and Taghizadeh [60] referred to algorithms that alter the phase constraint as iterative Fourier transform algorithms (IFTAs), and to those that alter output constraint as modified Gerchberg-Saxton (mGS) algorithms in the design techniques for diffractive optical elements and the application to fabre-coupling problems.

Mihailescu et al. [61] used GS algorithm implemented in Matlab CGH and DH (digital holography) with different sets of constraints in the input plane (hologram plane) and the output plane (object plane).

The quality of the numerically generated image can define by calculating the parameters:

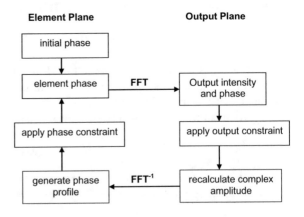

**Fig. 5.25** A schematic representation of the Gerchberg-Saxton algorithm

The mean square error [62]: MSE as the difference between the calculated intensity and desired one from initial image file;

The diffraction efficiency: as the ratio between the intensity in the signal window and the whole diffracted intensity;

The contrast in a given image: $\frac{Intens_{max} - Intens_{min}}{Intens_{max} + Intens_{min}}$, where $Intens_{max}$ and $Intens_{min}$ are the maxim and minimum values from all pixel of the intensity image, respectively [61, 63].

Other type of hologram is dot matrix holograms, which often are used for both security and decorative purposes. They are designed also on computer. Dot-matrix holograms consist of millions of tiny diffraction gratings, or "holopixels", oriented at different angles and arranged in a two-dimensional array. When illuminated with white light the holopixels break up the light into a spectrum of colors and redirect the light at various angles to form a kinetic hologram image [64, 65].

**E-Beam Lithography**

E-beam lithography (EBL) is the most commonly used technique for nanolithography in this kind of configurations. The process starts with the resist deposition by spin coating, electro beam exposure and resist development. After this, metal deposition and resist lift off defines the mask for the dry etching that transfer the pattern to the structural layer. The end of the process is the release of the structure by wet under etching of the sacrificial layer (Fig. 5.26).

The use of EBL for pattering represents many advantages that provide an ideal lithographic platform for the MEMS (NEMS) fabrication. Its high resolution is capable of defining features down to the nanoscale if needed. More important, the pattern design flexibility may be very convenient for prototyping, since in this device design optimization is often essential. Direct writing EBL may be limited for a rather low throughput, consequence of the serial addressing of the beam [66].

When fabricating such holograms it is necessary to verify that the CGH will create the desired wavefront. However it is not possible to check independently since the wavefront is aspheric. It is thus necessary to validate the fabrication

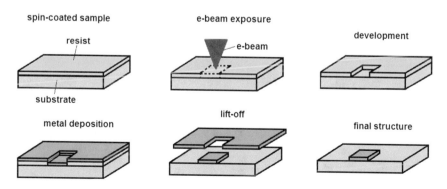

**Fig. 5.26** E-beam lithography process

process itself to insure accuracy of the hologram. State of the art laser beam and electron-beam lithography technology allow CGHs to be directly written at their finished size, which eliminate the photo-reduction process. The main sources of CGH errors made by direct writing can be divided into errors of the CGH structure (simulation, encoding, writing, and etching) and other errors (substrate figure, alignment, and optical test setup). The actual writing of the CGH pattern is the most critical fabrication step. The writing time can be from a few hours for laser writers to a few tens hours for e-beam writers. In spite of the long exposure process the absolute accuracy of positioning of writing beam in whole field of CGH must be not more than parts of micron [34].

## 5.2.1  The Creation and Formation of the Periodical Microstructure on the Basis of Computer Generated Hologram

Holograms are important in spheres, like scientific research, medicine, commerce industry etc. In mass production it is necessary to use high-quality originals in order to produce high-quality replicas. In order to produce high-quality replicas, new methods of production and development were created. Computer generated hologram (CGH) is one of such a techniques.

Computer-generated hologram is described mathematically by computing the information of phase and amplitude of the wave propagation produced by an object. CGHs are used in many applications, such as diffractive-optical elements for storage of digital data and images [67], precise interferometric measurements [68], pattern recognition [69], data encryption [70] and three-dimensional displays [71]. One of the advantages over conventional holograms, produced by optical means, is that the object used for recording CGH holograms does not necessarily exist, i.e. it may be described mathematically.

The picture and the created computer-generated hologram (CGH) in this work were analyzed in the plane of rectangular coordinates [72, 73]. The rectangular coordinates for initial picture were selected as shown in Fig. 5.27.

The parameters can be represented as:

$$\Delta x = \frac{1}{2X'}; \quad \Delta y = \frac{1}{2Y'}; \quad M = \frac{X}{\Delta x} = 2X'X; \quad N = \frac{Y}{\Delta y} = 2Y'Y.$$

The quantity of points in the picture is equal to MN. Switching to the new coordinates in the hologram (Fig. 5.28) and specifying that $X = Y = 1$, we obtain:

$$X' = \frac{M}{2}; \quad Y' = \frac{N}{2}; \quad \Delta x = \frac{1}{M}; \quad \Delta y = \frac{1}{N}$$

**Fig. 5.27** Picture plane

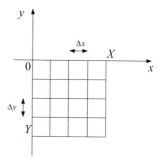

**Fig. 5.28** Hologram plane

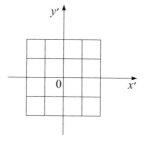

In the picture, coordinates of points are written as:

$$x_m = m\Delta x = \frac{m}{M}, \quad m = 0, \ldots, M;$$
$$y_m = n\Delta y = \frac{n}{N}, \quad n = 0, \ldots, N.$$

The number of hologram units is selected in the way, that unambiguous match would be ensured between the picture and discrete Fourier transformation, which is described in the hologram. The number of points in the plane will be equal to MN as well. This follows from the point that the system which is composed from MN points, the full system is composed from the functions of trigonometry which rate is:

$$X' = 0, \pm 1, \ldots, \frac{M}{2} - 1, \frac{M}{2};$$
$$Y' = 0, \pm 1, \ldots, \frac{N}{2} - 1, \frac{N}{2}.$$

The intensity of points of the hologram is specified in the coordinates (p, q), where the Fourier transformation of the function $h_{mn}$ is performed. Function $h_{mn}$ denotes the intensity of the point in the picture [72, 73]:

$$H_{kl} = \sum_{m=0}^{M-1} \sum_{n=0}^{N-1} h_{mn} e^{-i2\pi\left(k\frac{m}{M} + l\frac{n}{N}\right)}. \tag{5.10}$$

The process of formation of digital Fourier hologram was implemented by means of the program developed with MATLAB. The emblem of Kaunas University of Technology (KTU) was selected for experiments. First of all, coloured or grayscale version of the initial picture (Fig. 5.29a) was transformed into the black-white image (Fig. 5.29b). Subsequently the image was magnified four times, by placing original black-white picture (Fig. 5.29b) in the first quarter. This transformation was performed in order to eliminate corruption of the CGH after the reconstruction process. And finally CGH was created by applying Fourier transformation.

Computer generated hologram (Fig. 5.30) was then checked by applying the inverse Fourier transformation (Fig. 5.31). The program was developed with MATLAB for the purpose of reconstruction of hologram. It is necessary to mention that this hologram can be reconstructed by using any image display software, which has active FFT function. It is evident that in the reconstructed CGH only border-lines of the emblem were recorded.

Simulated CGH was fabricated using e-beam lithography. The formation of the surface relief procedure was performed on 300 μm-thick, $10 \times 10$ mm$^2$ silicon wafers, which were coated with 1.4 μm-thick polymethylmethacrylate (PMMA) layer. Specimen was exposed by using electronic lithography system "e-Line" (Raith GmbH) with $100 \times 100$ mm movement of laser interferometer-controlled stage and the "Gemini" electronic-optic system (Carl Zeiss). Main settings of the exposure:

- Accelerating voltage of the electron beam—20 kV;
- Electron beam current in the sample—0.3 nA;
- Aperture of electronic optical system—30 μm;
- Beam diameter of a focal plane—10 nm;
- Nominal exposing dose—36 μC/cm$^2$;

**Fig. 5.29** Initial grayscale (**a**), and black-white (**b**) images of the emblem of KTU

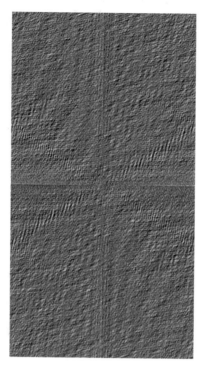

**Fig. 5.30** Computer-generated hologram

- Exposing step in x and y axis direction—40 nm;
- Exposure field size—100 μm.

Raster image of the black-white KTU emblem was converted using "3Lith" software (Raith GmbH) into vector format, which is better for exposition (GDSII format). GDSII image consists of squares, where lengths of sides indicate the Step parameters in conversion script.

The exposed plate was developed for 3 min in methyl ethyl ketone (MIBK) solution (MicroChem) at 20 °C. After the process of development, the sample was washed for 30 s in isopropyl alcohol and dried at room temperature with the flow of nitrogen gas.

For the control of micro and nano relief of 3D CGH produced in PMMA an atomic force microscope "EasyScan 2" (Nanosurf) (Fig. 5.32) was used in static constant force mode, using $2 \times 450 \times 50$ mm aluminium coated silicon probe (Nano World), with 13 kHz resonant frequency and 0.2 N/m force constant.

**Experimental Results**

Selected nominal dose was tested in four expositions by exposing a part of $200 \times 200$ μm image. Typical doses in this exposition were from 24.6 to 43.2 μC/cm$^2$, by 7.2 μC/cm$^2$. After the development of exposed CGH the relief was

**Fig. 5.31** Reconstructed
CGH

**Fig. 5.32** Scanning of CGH
surface using atomic force
microscopy

tested with atomic force microscope. Four different actual dose levels can be
determined from the topographical view (Fig. 5.33). The difference between the
lowest and the highest level of the topography is approximately 1200 nm. The
difference between the first and the second topographical levels constitutes 400 nm,
between the second and the third levels—450 nm, and between the third and the
fourth—350 nm. Differences are sufficient however they are not distributed evenly
and could have influence on the quality of reconstructed hologram.

**(a)**

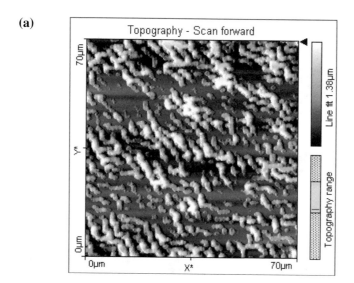

**(b)**

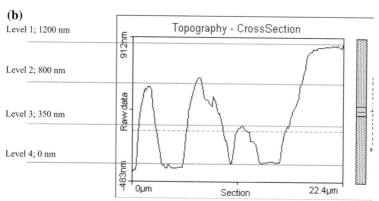

**Fig. 5.33** Topographical view of exposed specimen: **a** two dimensional topographical view, **b** the profile with obtained average topography levels, corresponding to the different actual doses

3D relief of formed CGH from PMMA was transferred to polymer using UV light hardening process (Fig. 5.34). UV light hardening replication was performed using commercial photopolymer (acrylic trimethylolpropane ethoxylate), PET substrate and custom-built technological device (T = 20 °C, irradiation distance—10 cm, UV light source DRT-230: $\lambda$ = 360 nm, I = 10,000 lx). Formed 3D relief thereby can be used as master for many times to replicate 3D structure.

Final product of CGH was tested visually with the laser ($\lambda$ = 633 nm). Photo of the reconstructed image is presented in Fig. 5.35. The quality of reconstructed hologram can be increased by changing exposure doses (depth of structure

**(a)**          **(b)**

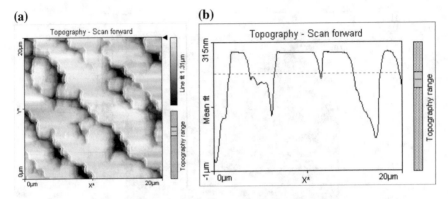

**Fig. 5.34** Topographical view (**a**) and profile (**b**) of the replicated structure in polymer

**Fig. 5.35** CGH
reconstructed by using laser
of $\lambda = 633$ nm

produced in the PMMA) in e-beam lithography and increasing number of grayscale
levels for CGH.

## 5.2.2   Gerchberg-Saxton Algorithm for Design
of Computer Generated Hologram

The Gerchberg-Saxton (GS) algorithm is an iterative Fourier-transform-based
algorithm, which calculates the phase required at the hologram plane to produce a

predefined intensity distribution at the focal plane. Unlike the gratings and lenses approach in which the phase between the traps is fixed, this algorithm provides phase freedom by iteratively optimizing both, the hologram and the image plane phase values. Because the beam shaping is limited to the focal plane, only 2D intensity patterns can be generated. A predefined intensity pattern, $I_d(x_i, y_i)$, can be anything from a single dot to a completely arbitrary distribution. The final goal is to find the phase at the hologram plane so that:

$$I_d = FFT\left(e^{\phi(x_h, y_h)}\right)^2 \tag{5.11}$$

(here FFT—fast Fourier transformation) which results in the desired pattern being transferred to the image plane. The algorithm is being initialized by assigning a random phase, $\phi_r$, and unit amplitude to the hologram plane. The first step of the algorithm is given by:

$$u_{h,1} = e^{i\phi_r}. \tag{5.12}$$

This field is then propagated to the image plane by taking its Fourier transform. This is done during each of n iterations as:

$$u_{n,i} = FFT\left(u_{h,n}\right) \tag{5.13}$$

Later on the phase from the resulting complex field at the image plane is retained, and the amplitude is replaced with amplitude, derived from the desired intensity:

$$\phi_{i,n} = \arg(u_{i,n}) \tag{5.14}$$

$$u_{i,n}^* = \sqrt{I_d}\, e^{i\phi_{i,n}} \tag{5.15}$$

By taking the inverse Fourier transform of Eq. (5.15) the field is propagated back to the hologram plane:

$$u_{h,n}^* = FFT^{-1}\left(u_{i,n}^*\right) \tag{5.16}$$

And finally, the phase is retained at the hologram plane and the amplitude is replaced again with uniform constant amplitude:

$$\phi_{h,n+1} = \arg\left(u_{h,n}^*\right) \tag{5.17}$$

$$u_{h,n+1} = e^{i\phi_{h,n+1}} \tag{5.18}$$

This completes one iteration giving a phase approximation that, when transformed, approximates the desired intensity. The algorithm quickly converges after completing few iterations, producing the desired phase $\phi_h = \arg(u_h)$. Moreover, as it was mentioned previously, the algorithm results in a hologram, which produces two dimensional intensity distribution or pattern [53, 74].

The process of formation of digital Fourier GS hologram was implemented by means of the program developed with MATLAB. Like in Sect. 5.2.1, the same logo of Kaunas University of Technology (KTU) was selected for experimental research (Fig. 5.36). CGH was created by applying Fourier transformation with GS algorithm (Fig. 5.37). The algorithm was run for 100 iterations.

Generated hologram (Fig. 5.37) was checked using the inverse Fourier transformation (Fig. 5.38). A program was developed with MATLAB for the purpose of reconstruction of the hologram. In the holographic map (Fig. 5.37) black color means particular place (point, mark), which is not displayed. Whereas white color means the maximum doze (the map has 8 levels).

### Experimental Process and Results

E-beam lithography was used for the formation of CGH into multilayer structure polymethyl methacrylate (PMMA)—silicon (Si) (Fig. 5.39). The size of the CGH is $2.048 \times 2.048$ mm (or $1024 \times 1024$ pixels).

The final product of CGH was tested optically. Picture of the reconstructed image is obtained from the metalized array, presented in Fig. 5.40 (laser wavelength $\lambda = 633$ nm).

### Comparison of Fourier and GS Algorithm Results

CGH made with both methods was compared optically: calculating diffractive efficiency and evaluating distribution of the diffraction maxima.

**Fig. 5.36** Initial grayscale image of the emblem of KTU

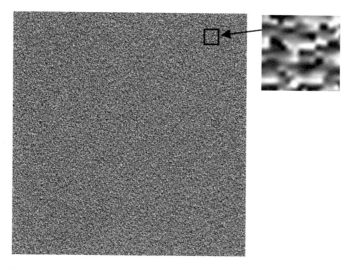

**Fig. 5.37**  Computer-generated hologram

**Fig. 5.38**  Reconstructed CGH

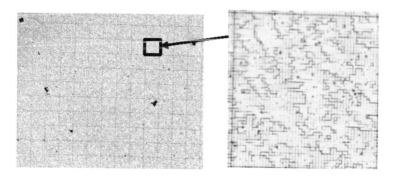

**Fig. 5.39**  Photo of the CGH exposed on PMMA

**Fig. 5.40** Fourier GS CGH
reconstructed using laser of
$\lambda = 633$ nm

Optical result is much better in the case, when CGH is created using GS algorithm than using only Fourier transformation (Figs. 5.35 and 5.40).

Using the Fourier transformation, the reconstructed CGH has only outlines of pictures and two grey-scale levels: black and white. When we modified the first model with GS algorithm, the reconstructed CGH had grey-scale levels and full pictures.

Diffraction efficiency of CGH increased approximately 6 times (from 8% in Fourier CGH to 45% in Fourier GS CGH). Differences in distribution of diffraction maximums were observed (Figs. 5.41 and 5.42) as well.

## 5.3 High-Frequency Excitation for Thermal Imprint of Microstructures into a Polymer

### 5.3.1 Methods of Microstructure Replication

Well-known conventional technology such as injection molding, injection compression molding and hot embossing have been extensively used at the micro-replication scale. Hot embossing is now becoming a promising manufacturing process, which is well suited for producing dedicated microstructures with high aspect ratios and small distortions [75, 76].

The injection molding process involves the injection of a melt polymer into a mold where the melt cools and solidifies to form a plastic part. It is generally a three phase process including filling, packing and cooling phases. After the cavity becomes stable, the product is ejected from the mold [77] (Fig. 5.43).

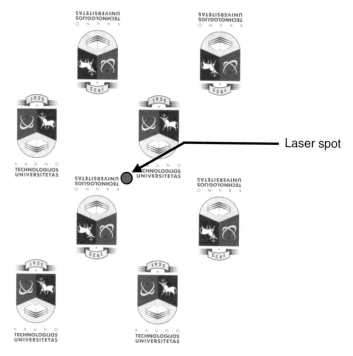

**Fig. 5.41** Distribution of the diffraction maximums of the reconstructed Fourier CGH

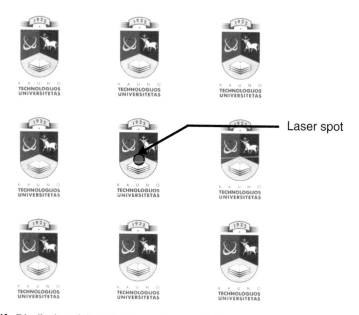

**Fig. 5.42** Distribution of the diffraction maximums of the reconstructed Fourier GS CGH

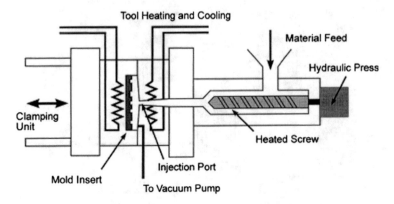

**Fig. 5.43** Schematic diagram of an injection molding machine [78]

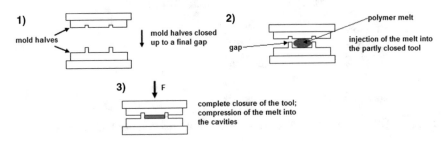

**Fig. 5.44** Schematic view of the injection compression molding process [76]

Injection compression molding is a combination of injection molding and hot embossing (Fig. 5.44). First, a volume of polymer melt corresponding to the volume of the molded part is injected into a mold that is not completely closed. Because of the gap between the mold halves the injection can be done by lower pressure, resulting in a reduction of shear velocity and shear stress of the polymer melt. After the injection step the mold halves are closed by a compression step. This compression step is split into a velocity-controlled motion and finally, if the desired clamp force is achieved, in a force-controlled holding of the final press force over a defined cooling time [76].

As shown in Table 5.4, hot embossing provides several advantages compared to injection molding and other processes, such as relatively low costs for embossing tools, a simple process, and high replication accuracy for small features. Therefore it was selected for microstructures replication.

**Table 5.4** Comparison between molding methods [79]

| | Liquid resin molding | Injection molding | Hot embossing |
|---|---|---|---|
| Mold inserts | Any molds | Metal molds (Silicon molds suitable for prototyping) | Metal and silicon molds |
| Feature size | No limit | Good for small features with low aspect ratio, or large features with high aspect ratio<br>Good for 3-D dimensional features | Good for small features<br>Difficult for high aspect ratio<br>Difficult for multiple depth<br>Planar features only |
| Material | Liquid resins (thermosets provide high chemical resistance) | Mainly low molecular weight thermoplastics | Low and high molecular weight thermoplastics |
| Processing | Simple<br>Easy mold filling<br>Closed mold process<br>Long cycle time<br>Mold release problem for some resins | Short cycle time (sec ~ min)<br>Closed mold process<br>High automation | Simple<br>Medium cycle time (min)<br>Potential for continuous production<br>Open mold process |
| Replication accuracy | Less dimensional control (polymerization shrinkage) | Excellent dimensional control | Less dimensional control |
| Part quality | Low molded-in stresses<br>Contamination (resin residue) | High stress on mold insert<br>High molded-in stresses | High molded-in stresses |
| Cost | Low tooling cost (except for RIM)<br>For prototyping and low volume production | High tooling cost<br>For large volume production | Low tooling cost<br>For low and medium volume production |

### 5.3.1.1  Hot Embossing Process

Hot embossing [80] is a manufacturing technique used for a wide number of applications from macro to nanometer scales. Hot embossing is a technique of imprinting microstructures on a substrate using a master mold. It mainly consists of the following steps. First, the polymer heated above its glass transition temperature $(T_g)$. Secondly, the polymer is stamped with the desired imprint at this temperature (above $T_g$) where it behaves more fluid-like. Also, the force required to deform the polymer at this temperature is much smaller than that required at temperatures below the glass transition $(T_g)$ of the polymer. Thirdly, the polymer is cooled below its $T_g$ and the mold is removed [81].

Hot embossing can be cyclic and continuous process [82]. Li et al. [83] made a series of experiments to investigate the processing of micro-components by hot embossing. The results demonstrated that the replication accuracy strongly depends on the processing conditions: on the processing temperature and pressure.

It is essential to analyze the different parameters associated with the final quality control of the replicas obtained by hot embossing. Recent works describe the influence of polymer materials (viscosity, forming temperature etc.), those depending on the mould and related parameters (material, geometry, physical and thermomechanical properties, surface states and temperature) and those describing the hot embossing process (temperature and heating time, holding and cooling time, applied pressure, demolding temperature, etc.) are important [84, 85].

In the hot embossing step, non-uniform pressure distribution may result in pattern height differences [86]. The different thermal expansion coefficient may appear a large thermal stress. Therefore, it is possible fracture of the polymer micro structure during the cooling step. The adhesion between the polymer and mold in the demolding step may influence final pattern defects [87].

When the imprint pressure is not high enough, much longer holding time are needed to provide complete pattern filling, which will result in long cycle times and low embossing efficiency. Although high pressure can improve the embossing efficiency, too high pressure can damage the friable mold as silicon. The functionality micro structures with special use may be affected; material use for photoelectric detector, for example, will be damaged when the stress exceeds 6 MPa [88].

For the thermal-plastic material, the mechanical behavior is largely influenced by the temperature. When the temperature is too low, the polymer will lack flowability, which will result in a high amount of recovery and large distortions after demolding. When the imprint temperature is too high, the polymer molecule can be broken and many defects develop in the imprint area. Hu [88] reported that there were many bubbles in the PMMA substrate after embossing when the imprint temperature reached 280 °C. On the other hand, to avoid significant material out sequence at the die-to-die interface, a stable deformation of the polymer at relatively low temperatures is desired [89].

Many authors highlight the following hot imprint process problems:

- Filling ratio of microstructure [90];
- Non-uniform mold imprint [91];
- Adhesion between mold and polymer [92];
- Surface roughness [93];
- long cycle time [94].

However, only hot imprint process parameters optimization is not enough. It is need to take into account the different technological equipment, materials and etc. This requires new methods and are as for improvement to achieve better quality of replicas. These problems can be solved using high frequency vibrations [95, 91].

### 5.3.1.2 Ultrasonic Hot Embossing

High frequency vibrations are widely used in different applications and technologies. Capillary waves are used for droplet formation on a vibrating surface in [96]. Surface acoustic waves are used to concentrate bioparticle suspensions [97], to control the temperature of liquid droplets [98], to generate solitary pulses and fracture [99], to produce regular, long-range, spatially ordered polymer patterns without requiring the use of physical or chemical templating [100]. Ultrasonic motors are used to drive fluids [101], to assist cardiac compression devices [102], to control electro-rheological fluids [103].

The micro patterning of polymers by ultrasound was first described for melting and molding of powders [104, 105]. The development of micro patterning of polymer plates started [95] and in 2008 there was the first publication on a micro system fabricated by ultrasonic hot embossing [106].

Ultrasonic vibration [107, 108] has been widely used in industry in the welding and joining of thermoplastics that have a low softening temperature. The equipment required for ultrasonic vibration hot embossing can include a fixture for holding the parts, a vibration horn, an electromechanical transducer to drive the horn, a high frequency power supply, and a cycle timer. During ultrasonic vibration, high frequency (typically 10–40 kHz) low amplitude (typically 1–25 μm) mechanical vibrations are applied to the parts. This results in cyclical deformation of the parts and of any surface roughness. The ultrasonic energy is converted into heat through the intermolecular friction within the thermoplastics. The generated heat, which is highest at the surface between the master mold and the plate due to asperities, is sufficiently high to melt thermoplastics and cause the melt to flow to fill the interface. Ultrasonic vibration heating can provide an effective way of heat generation to hot emboss the precise structure onto the surface of a large plate [95].

Liu and Dung [95] use ultrasonic vibration as a heat generator for hot embossing. They investigated the replication capacity of ultrasonic-heating embossing of both amorphous and semicrystalline plastic plates; examined the effects of various ultrasonic vibration parameters on the contour of replicated structure; identified the relative significance of all these parameters on molded part quality. The experimental results showed that ultrasonic vibrated hot embossing could provide an effective way of molding precise-structures onto polymeric plates with good replicability. Amorphous materials exhibited better replicability than semicrystalline materials in ultrasonic vibrated embossed plates. The replicability of embossed plates decreased with plate thickness. One can improve the replicability of an ultrasonic vibration hot embossing plate by: increasing start pressure, amplitude of vibration time and vibration pressure, and hold pressure and hold time. In addition, the replicability of embossed plates increases with the energy input. However, if the energy input is too high, the replicability decreases mainly due to overmelt of the plate.

Mekaharu et al. [91] succeeded in precise replication of patterns by impressing ultrasonic vibration besides heat and the loading force on the micro hot embossing (MHE). They used a Ni electroformed mold with micro patterns with seven sizes of entrance from 100 $\mu m^2$ to 1.2 $mm^2$. The experiment was executed on a polycarbonate (PC) sheet while impressing longitudinal wave of ultrasonic vibration at maximum amplitude of 1.8 $\mu m$. The effect that bubble defect in the pattern was diminished or completely disappeared was observed by impressing ultrasonic vibration. Moreover, it was clarified that ultrasonic vibration assisted softened PC to move to the center of the pattern area in a mold. They found that the effect of ultrasonic vibration was well pronounced at low contact force. Perhaps very high contact force may be obstructing the progression of ultrasonic vibration.

In this chapter the piezo ceramics are used as generator of vibration. A detail of their properties and applications is presented in the next section.

## 5.3.2  Materials, Experimental Setup and Methodology

In this section materials used in mechanical hot imprint process, their properties, devices used in microstructure quality assessment, tools used to obtain the dynamical properties of vibroplatform and methodology for the evaluation of the damping ratio are presented.

### Polycarbonate
Experimental studies of thermal imprint process were done using polycarbonate. The physical, chemical and thermal properties of the polycarbonate are shown in Tables 5.5, 5.6 and 5.7.

### Piezo Elements
In the experiment ring form piezo ceramic discs PZT-4 were used. They are recommended for their high resistance to depolarization and low dielectric losses under high electric drive. Their high resistance to depolarization under mechanical stress makes them suitable for use in deep-submersion acoustic transducers and as the active element in electrical power generating systems. The PZT-4 material properties are shown in Table 5.8 [109].

**Table 5.5**  Physical properties

| Density (ρ) | 1.20–1.22 g/cm$^3$ |
|---|---|
| Refractive index (n) | 1.584–1.586 |
| Water absorption—equilibrium (ASTM) | 0.16–0.35% |
| Water absorption—over 24 h | 0.1% |
| Light transmittance | 88% |

**Table 5.6**  Chemical properties

| Young's modulus (E) | 2.0–2.4 GPa |
|---|---|
| Tensile strength (σt) | 55–75 MPa |
| Compressive strength (σc) | >80 MPa |
| Poisson's ratio (ν) | 0.37 |
| Coefficient of friction (μ) | 0.31 |
| Speed of sound | 2270 m/s |

**Table 5.7**  Thermal properties

| Melting temperature (Tm) | 267 °C |
|---|---|
| Glass transition temperature (Tg) | 150 °C |
| Heat deflection temperature—10 kN | 145 °C |
| Heat deflection temperature 0.45 MPa | 140 °C |
| Heat deflection temperature—1.8 MPa | 128–138 °C |
| Upper working temperature | 115–130 °C |
| Lower working temperature | −40 °C |
| Linear thermal expansion coefficient (α) | 65–70 × 10$^{-6}$ K |
| Specific heat capacity (c) | 1.2–1.3 kJ/(kg K) |
| Thermal conductivity (k) 23 °C | 0.19–0.22 W/(m K) |
| Thermal diffusivity (a) 25 °C | 0.144 mm$^2$/s |

**Table 5.8**  Properties of PZT-4

| General | | | |
|---|---|---|---|
| Density $(10^3 \text{ kg/m}^3)$ | 7.5 | | |
| Curie temperature (K) | 601 | | |
| Elastic constants | | Piezoelectric constants | |
| $c_{11}^E$ (GPa) | 139 | $e_{31}$ (C/m$^2$) | −5.2 |
| $C_{33}^E$ (GPa) | 115 | $e_{33}$ (C/m$^2$) | 15.1 |
| $c_{12}^E$ (GPa) | 77.8 | $e_{15}$ (C/m$^2$) | 12.7 |
| $c_{13}^E$ (GPa) | 74.3 | Dielectric constants | |
| $c_{44}^E$ (GPa) | 25.6 | $\varepsilon_{11}^T$ (10$^{-9}$ F/m) | 6.461 |
| $c_{66}^E$ (GPa) | 30.6 | $\varepsilon_{33}^T$ (10$^{-9}$ F/m) | 5.620 |

## PRISM Setup for Obtaining of Dynamical Properties

A number of experimental studies are needed in order to ensure high dynamic accuracy of operation of optical scanners. In most cases the exciting frequencies are quite high, and the amplitudes corresponding to them are measured in micrometers.

Therefore the holographic method can be effectively applied for the visual representation of dynamic processes, taking place in the waveguide of the optical scanner. The most effective method for studying these dynamic processes is the method of digital holographic interferometry.

No contacting holographic measure-system PRISM (produced in USA by company HYTEC) was used in order to establish dynamical properties of piezoceramics and vibroplatforms.

PRISM system combines all the necessary equipment for deformation and vibration measurement of most materials in a small lightweight system. A standard system includes holography and computer systems integrated with proprietary state of the art software. The main parts of the PRISM system are presented in Fig. 5.45.

The PRISM system (Table 5.9), shown in Fig. 5.45 is a two beam speckle pattern interferometer. In the study green (532 nm, 20 mV) laser is used. The laser beam, directed at the object is called the object beam, the other beam, which goes directly to the camera, is the reference beam. Laser light is being scattered from the object and collected by the camera lens, which images the object onto the CCD camera sensors. The image is then sent from the camera to a computer, analyzed with program PRISM DAQ and dynamic processes that take place in the sample can be observed in the monitor (Fig. 5.46).

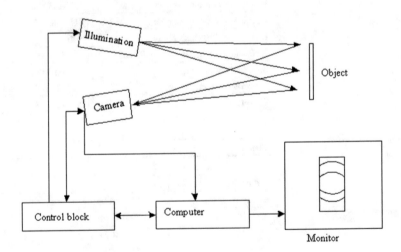

**Fig. 5.45** PRISM systems optical setup

| **Table 5.9** Specification of PRISM system | | |
|---|---|---|
| | Measurement sensitivity | <20 nm |
| | Dynamic range | 10 μm |
| | Measurement range | >100 μm |
| | Largest part size | 1 m diameter |
| | Working distance | >1/4 m |
| | Data acquisition rate | 30 Hz |
| | Laser | 20 mV |

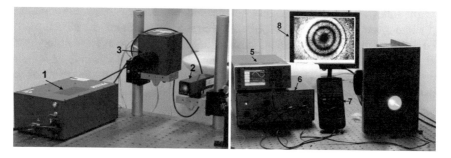

**Fig. 5.46** PRISM system: *1*—control block, *2*—illumination head of the object, *3*—video head, *4* —vibroplatform, *5*—amplifier, *6*—generator, *7*—tester, *8*—interference fringe

The best quality interfering pattern, in the control block and camera is replaced. The ratio between the object and the reference beams—1:2.4.

It is necessary to make assumption that time varying displacement is along z-axis, which, in the holographic arrangement is used. It is along the line of sight between object and observer. The displacement is a periodic function of time and the development is simplified if the displacement is allowed to vary only with x and time, it is presumed that $Z(x)\sin(\omega t)$. If $\phi(x, y)$ is the resting phase distribution, then the object complex amplitude at the film plane is

$$U_O = A(x, y)e^{i\left[\phi(x,y) + \frac{4\pi}{\lambda}Z(x)\sin\omega t\right]}; \tag{5.19}$$

where $\lambda$ is the wavelength of the laser.

The time-average hologram is recorded with object beam and reference beam for a time T, which is longer than several periods of the vibration. The reconstructed object wave has complex amplitude that is proportional to the time average of the $U_O$ over time T, which is:

$$U_{Oav} = A_{av}(x, y)\frac{1}{T}\int_0^T e^{i\left[\phi(x,y) + \frac{4\pi}{\lambda}Z(x)\sin\omega t\right]}dt = J_0\left[\left(\frac{4\pi}{\lambda}\right)Z(x)\right]. \tag{5.20}$$

The $J_0$ is the zero-order Bessel function. The irradiance is calculated as following:

$$I(x, y) = A^2(x, y)J_0^2\left[\left(\frac{4\pi}{\lambda}\right)Z(x)\right]. \tag{5.21}$$

In this case, the image has been superimposed on a system of fringes, which correspond to the minima of the square of the zero-order Bessel function [110].

The following procedure should be completed in order to calculate the amplitude of vibrations of the plate shaped object, which is mounted tightly by its end on the fixture. The fringes of the vibrating plate, obtained by the time-average holographic interferometry are schematically shown in Fig. 5.45. The point P is in the middle of the second dark fringe. The centers of dark fringes coincide with the points of the plate, where the amplitude of vibrations Z(x) is such, that the Bessel function obtains zero value.

The higher-order (higher then 20) zeros of the Bessel function $(\xi_n)$ are set almost equally and can be depicted by the following equation [110]:

$$\xi_n = \left(n - \frac{1}{4}\right)\pi + \frac{1}{8}\left[\left(n - \frac{1}{4}\right)\pi\right]^{-1} \qquad (5.22)$$

Then amplitude of the vibrations (Z) in the point P can be determined by the following equation:

$$\xi_n = \frac{4\pi}{\lambda}Z(P). \qquad (5.23)$$

Hot imprint device

Joint-Stock Company "Holography industry" is one of the leading enterprises, which manufactures high-quality polygraphic production. According to the company the high scientific and technical achievements in the MMI was designed hot imprint device. This hydraulic hot imprint device was used in hot imprint experiment (Fig. 5.47).

The main parts of the hot imprint device are: hydraulic hold (1), gauge of pressure (2), mold horn (3), controlled stage (4), thermometer (5), dynamometer (6), and control block of temperature, time, and pressure (7). The specifications of parameters are shown in Table 5.10.

**Fig. 5.47** The device of hot imprint process

**Table 5.10** The specifications of hot imprint device

| Range of temperature | 20–200 °C |
|---|---|
| Range of pressure | 0–200 kg |
| Horn measurement | 2 cm × 2 cm |

**Fig. 5.48** The frequency generator

In the hot imprint process experiment frequency generator Г3-56/1 was used (Fig. 5.48).

**Tools for Quality Assessment of the Microstructure**

**Optical Microscope**

The optical microscope "NICON Eclipse LV 150" (Fig. 5.49) with CCD camera was used for investigations. Microscope can magnify the view by 25×, 50×, 100×. Other parameters are: distance between oculars 47–82 mm, rotational angle 360°, mobility in X and Y axis, observation field 20 mm × 20 mm, illumination by halogen lamp (12 V, 50 W), resolution 2 megapixels, color sensor resolution 1600 × 1200, dynamic range >60 dB, power 2.5 W, integration time 84 μs to 3 s.

As shown in Fig. 5.49 the main components of the digital optical microscope are: 1—trinocular tube, 2—LV-UEPI2 Illuminator, 3—CFI LU plan flour objectives, 4—stage, 5—CCD camera, 6—specimen, 7—view of specimen. Data are recorded and processed by computer and analyzed by software program InfinityCapture.

**Diffractometer**

Optical properties of the replica were evaluated using non-destructive optical method. Laser diffractometer (He–Ne, $\lambda = 632.8$ nm, 50 mW) was used in order to register reflection or spectrum of transmitting diffraction (Fig. 5.50). Diffraction efficiencies of diffracted light were measured by a photodiode in all maximums ($0, \pm1, \pm2$, and etc.) for different angles of incidence light with respect to the normal.

The measuring schema of diffraction efficiency is showed in Fig. 5.50. The main components are: sample (1), photodiode (2), tester (3) and maximum distribution (4).

The measurements were performed in 5 different points of each sample and mean values were calculated. The laser beam is directed into the sample's (periodic

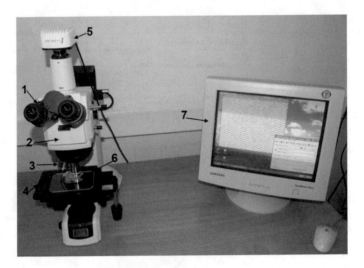

**Fig. 5.49** Optical microscope Nikon Eclipse LV 150

**Fig. 5.50** Difractomer and
measuring schema

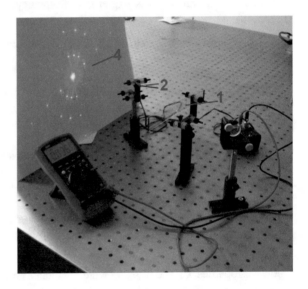

microstructure) surface. From transmitted and the reflected light's maximum goes
into photodiode and is registered with tester. Electrical current, which passes
through the photodiode, linearly depends on the lighting, so experimental results
can be compared without any additional calculations.

All periodical microstructures formed in optical materials are characterized by
relative diffraction efficiencies. Relative diffraction efficiency $RE_{i,j}$ is defined as
ratio of intensity of diffracted light $I_{i,j}$ to the i-th diffraction maximum and j-th
illumination angle with surface without microrelief divided from $I_j$—total

transmission or reflection of the intensity of diffraction maximum at the j-th illumination angle

$$RE_{i,j} = \frac{I_{i,j}}{I_j}. \tag{5.24}$$

$$I_j = \sum_i I_{i,j} \tag{5.25}$$

The relative efficiency allows eliminate the material optical properties of which periodic microstructure is made, and evaluate the geometry of the structure. This allows compare the results of various technological processes, where the periodic microstructure is made of different materials.

**Atomic Force Microscope**

Measurement of geometrical parameters of periodical structures was performed using atomic force microscope NANOTO P-206 (AFM) (Fig. 5.51).

The microscope consists of two parts: scanning part 1 and controller 2. AFM works at ambient temperature. The mechanism is able to move 20 mm and the step can be 2 nm. AFM has following modes: noncontact, contact and tapping. It depends on the probe contact with the surface, which is being investigated. AFM NANOTOP-206 specifications are presented in Table 5.11.

AFM recorded and statistically evaluated following sample's surface parameters (chosen on the basis of the standard ASMEB46.1-1995):

A     Maximum Height
$Z_{mean}$     Average Height;
$R_a$     Average Roughness
$R_q$     Root-Mean-Square Roughness. A higher $R_q$ value corresponds to a higher surface roughness;
$R_{sk}$     Skewness corresponds the surface profile variation from the centerline. When $R_{sk} < 0$—elements are below the centerline, $R_{sk} = 0$—uniform distribution of elements, $R_{sk} > 0$—dominate elements above the surface

**Fig. 5.51** Atomic force microscope (AFM) NANOTOP-206

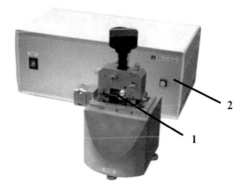

| Table 5.11 AFM specifications | Maximal scan field | To $12 \times 12$ μm |
|---|---|---|
| | Matrix of measurement | To $512 \times 512$ points |
| | Maximal high of measurement | 2.0 μm |
| | Lateral resolution | 2 nm |
| | Vertical resolution | 0.1–0.2 nm |

line. Surfaces with a positive $R_{sk}$ are characterized as having a sharp bumps with a height greater than average height. Surfaces with negative $R_{sk}$ are porous and have deep bumps. In rare cases, surfaces have $R_{sk} = 0$

All values, obtained using the AFM image, are processed using software SurfaceView (version 2.0).

**Oscilloscope**
Resonant frequency and damping coefficient were determined using bump test system. It consists mainly from laser displacement meter and computer oscilloscope PicoScope 3424 (Fig. 5.52). The device writes into computer memory the multi frequency electrical signal and performs the spectral analysis (four signals are recorded instantly). Signals are received from various sensors (vibration, pressure, force, acoustic, etc.). The data is being recorded in computer and presented graphically using the software PicoScope R5.16.2. Specifications are given in Table 5.12.

**Fig. 5.52** The bump test system: vibroplatform (*1*), laser displacement sensor LK-G82 (company KEYENCE) (*2*), oscilloscope PicoScope 3424 (*3*), DC supply for the displacement sensor controller (*4*), controller LK-G3001PV of displacement sensor (*5*), computer (*6*)

**Table 5.12** Specification of oscilloscope "PicoScope 3424"

| Bandwidth | 10 MHz |
|---|---|
| Channels | 4 |
| Vertical resolution | 12 bits |
| Dynamic range | 72 dB |
| Input ranges (full scale) | ±20 mV to ±20 V in 10 ranges |
| Input characteristics | 1 MΩ in parallel with 20 pF |
| Input type | Single-ended, BNC connector |
| Timebase ranges | 500–50 s/div in 25 ranges |
| Timebase accuracy | 100 ppm |
| Frequency range | DC to 10 MHz |
| Display modes | Magnitude, peak hold, average |
| Window types | Blackman, Gaussian, triangular, Hamming, Hann, Blackman-Harris, flat-top, rectangular |

The sample (1) is exposed to low mechanical impulses (in the center of the specimen). Excited mechanical vibrations are registered by laser sensor (2) and transferred to the oscilloscope (3). The signal is then processed using special software and PicoScope, and the results are represented on the display of computer (6).

Damping measurement requires a dynamic test. A record of the response displacement of an underdamped system can be used in order to determine the damping ratio.

The logarithmic decrement can be calculated by following formula

$$\delta = \frac{1}{n} \ln\left(\frac{x(t)}{x(t+nT)}\right),\qquad (5.26)$$

where n is any integer number of successive (positive) peaks. The values x(t) and x (t + nT) are two successive peaks.

Damping ratio can be calculated by following formula

$$\varsigma = \frac{\delta}{\sqrt{4\pi^2 + \delta^2}}.\qquad (5.27)$$

### 5.3.3   Investigation of Mechanical Hot Imprint Process

In this section the experimental technology, which is used for the quality optimization of complex microstructure replicas based on high frequency vibration in the mechanical hot imprint process is presented.

Hot imprint is operation with high accuracy in replicating micro-features. The cost of the embossing tools is relatively low, due to this it is popular manufacturing process. The hot imprint experiments were performed with polycarbonate at different temperatures, pressures and imprint times. The comparative analysis of microstructure quality assessment (with and without the usage of high frequency vibration) is presented.

As it was discussed in Sect. 5.3.1, in order to obtain replicas of best quality, it is not enough to apply the available devices and optimization. High frequency vibrations are one of the measures, which could be used in order to optimize mechanical hot imprint process. The mechanical hot imprint process is a complex technological operation, because the state of the material changes. Firstly, the mold was preheated till proper temperature. Secondly, the microstructure was imprinted into polymer. Polymer is plastic, so after demolding it remains deformed.

The research problem: after the hot imprint process in replicated microstructure remains empty cavities not filled by polymer, which affects the overall quality of replication. Due to this reason, to the experimental hot imprint process scheme a piezo element which generates high-frequency vibrations was added. This results in proportional distribution of the polymer in the microstructure and improves filling ratio. In this way the quality of obtained replica is improved.

The experiment was performed in the laboratory of KTU MI using hot imprint device (Fig. 5.47). The main parts of the hot imprint device are presented in Fig. 5.53.

Flat embossing experiments are performed using a flat thermal pressure device (Fig. 5.47). The original construction secures a controlled pressure, force, temperature and the duration of exposure (P = 1–5 Atm, T = 140–200 °C, t = 1–15 s) to a polymer (mr-I 8020). The surface contour of the nickel mold is transferred to the thin polymer film, which is coated onto glass, during the process of replication. The thickness of polymer is 400 nm.

Simple experiments show that the quality of a replica is quite low after the hot imprint. Therefore posibility to improve the quality of replica using high frequency excitation of the polymer surface during the process of thermal imprint was investigated. The piezoelectric element PZT-4 was chosen as a source of high

**Fig. 5.53** The schematic diagram of the flat thermal embossing device: *1*—the coin, *2*—the master structure, *3*—glass coated by the polymer, *4*—the base

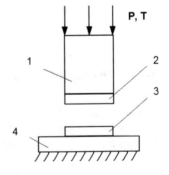

**Fig. 5.54** The schematic diagram of the thermal imprint experimental setup with high frequency excitation: *1*—the coin; *2*—polycarbonate; *3*—the piezoelectric element; *4*—the base; *5*—the vibrating platform

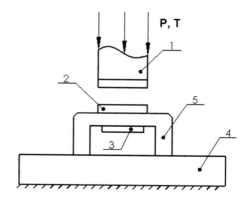

frequency vibrations. Dimensions of the element are as following 50 mm external diameter ring; 20 mm internal diameter and 3 mm thickness. The piezoelectric exciter was mounted under the platform, which holds the polymer in order to eliminate possible shortcuts or cracks caused by the pressure. The schematic diagram of the thermal imprint experimental setup with the piezoelectric element, which is mounted under the plate holding the polycarbonate, is shown in Fig. 5.54.

An aluminum cylinder with the top surface and a mounting hole in the side wall was chosen as a vibrating platform (Fig. 5.55a). The drawing of the vibrating platform with the mounted piezoelectric element is shown in Fig. 5.55b; the scheme of the electric circuit is presented in Fig. 5.56.

The material and geometrical parameters of the vibration platform were chosen according to the conditions of application (Figs. 5.55, 5.56 and 5.57)—the platform should sustain the pressure of 5 Atm and at the same time it should be flexible enough to transmit vibrations to the specimen.

### 5.3.3.1 Modeling and Experimental Research of Vibroplatform

The operating parameters of the vibration platform, which was used in order to improve the quality of replicas of complex microstructures during the mechanical hot imprint process (Fig. 5.54), are analyzed in this chapter. This vibroplatform was

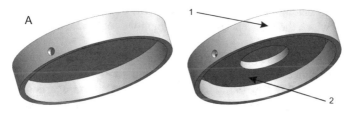

**Fig. 5.55** The 3D view of the vibration platform (**a**) and the vibration platform (*1*) with the piezoelectric element (*2*) attached (**b**)

**Fig. 5.56** The schematic
diagram of electric circuit of
the piezoelectric element

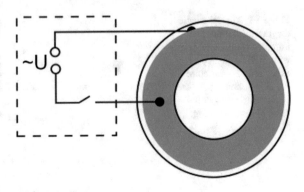

**Fig. 5.57** The drawing of the
vibration platform

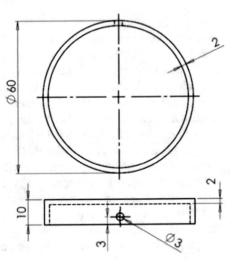

used as an additional tool in mechanical hot imprint process as a generator of high
frequency vibrations in order to enhance the filling of microstructure's gaps with
polymer. This analysis requires comparison between experimental and simulative
results. The created mathematical model of vibroplatform was implemented by
FEM using Comsol Multiphysics software. The vibroplatform's numerical analysis
was performed with and without the action of mechanical load. The damping ratio,
which is necessary for vibroplatform's numerical model, was obtained analytically.

In order to determine the damping coefficient of the vibroplatform, the bump
test, during which the transient oscillation graph was obtained, experimentally was
performed (Fig. 5.58).

Vibration amplitudes $x(t_0)$ and $x(t_n)$ were measured during time intervals t0 and
$t_n$. The logarithmic decrement δ and damping ratio ζ were calculated using
Eqs. (5.26) and (5.27). The bump test for each vibroplatform was done five times.
The δ and ζ were calculated by five different n: 1, 6, 10, 16, 21 in each graph. In
order to evaluate the resulting noise of the experiment, the average value of
damping ratio was calculated. The results are presented in Table 5.13.

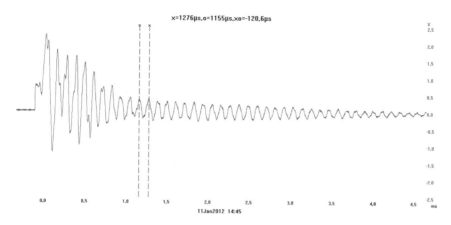

**Fig. 5.58** The transient oscillation of vibroplatform (2 mm thickness)

**Table 5.13** Damping ratios of three types vibroplatforms

| Thickness of vibroplatform (mm) | Average of logarithmic decrement | Average of damping ratio |
| --- | --- | --- |
| 1 | 0.4 | 0.06 |
| 2 | 0.3 | 0.05 |
| 3 | 0.13 | 0.02 |

In this chapter modeling of vibroplatform, which was used in hot imprint process as frequency generator, is presented.

An aluminum cylinder with the top surface and a mounting hole in the side wall was chosen as a vibrating platform. The drawing of the vibrating platform with the mounted piezoelectric element is shown in Fig. 5.59.

The applicability of the vibration platform in the process of hot embossing was analyzed numerically using finite element method. The dynamics of the platform (Fig. 5.59) was calculated using program COMSOL Multiphysics 3.5a.

There were analyzed vibroplatforms of three different thicknesses $h_{Al}$ (1, 2, and 3 mm), while thickness of piezo ceramic $h_{PZT}$ is 3 mm and total height of the vibroplatform h is 10 mm. Vibroplatform's bottom surface is fixed. Considering to the experimental results potential difference Q between the upper and lower piezo ceramic's surfaces varies from 5 to 150 V. The quarter-section of the computational scheme of vibroplatform is presented in Fig. 5.60.

The platform's material is aluminum and the disk is modeled as a piezo ceramic material PZT-4. The PZT-4 material's and aluminum's parameters were taken from Comsol Multiphysics material library. The tetrahedral quadratic was chosen as a mesh element. Fine mesh guarantees the convergence of the solution. The tetrahedral element (Fig. 5.61) is defined by ten nodes having three degrees of freedom at each node: displacement in the nodal x, y and z directions.

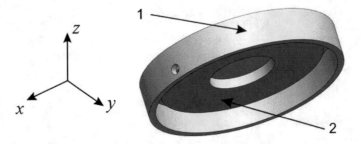

**Fig. 5.59** The 3D view of the vibration platform (*1*) with the piezoelectric element (*2*)

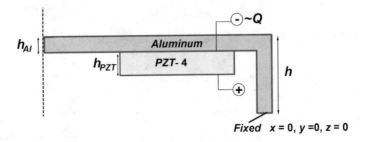

**Fig. 5.60** Computational scheme of vibroplatform

**Fig. 5.61** Tetrahedral finite element [111]

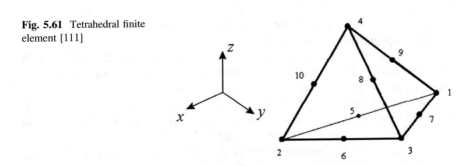

The material parameters for the piezoelectric material were specified: by selecting the stress-charge form based on the constitutive equation and entering the material data: the elasticity-matrix elements in the $c_E$ matrix, the piezoelectric coupling-matrix elements in the e matrix, and the relative permittivities in the $\varepsilon_{rS}$ matrix.

The material properties in this model were set considering that the polarization is in the z direction, which is a common orientation, according to the literature.

**Elasticity Matrix**

Defines the stress-strain relation matrix $c_E$

$$\sigma = c_E \varepsilon \tag{5.28}$$

where $\sigma$ is stress, and $\varepsilon$-strain.

$$c_E = \begin{bmatrix} 1.38999e11 & 7.78366e10 & 7.42836e10 & 0 & 0 & 0 \\ 7.78366e10 & 1.38999e10 & 7.42836e10 & 0 & 0 & 0 \\ 7.42836e10 & 7.42836e10 & 1.15412e10 & 0 & 0 & 0 \\ 0 & 0 & 0 & 2.5641e10 & 0 & 0 \\ 0 & 0 & 0 & 0 & 2.5641e10 & 0 \\ 0 & 0 & 0 & 0 & 0 & 3.0581e10 \end{bmatrix}$$

**Coupling Matrix**

Defines the piezo coupling matrix e, used in the stress-charge form of the constitutive equation

$$\sigma = c_E \varepsilon - e^T E, \tag{5.29}$$

where $\sigma$ is stress, $\varepsilon$-strain, and E is the electric field.

$$e = \begin{bmatrix} 0 & 0 & 0 & 0 & 12.7179 & 0 \\ 0 & 0 & 0 & 12.7179 & 0 & 0 \\ -5.20279 & -5.20279 & 15.0804 & 0 & 0 & 0 \end{bmatrix}$$

**Relative Permittivity**

The relative permittivity $\varepsilon_{rS}$ appears in the constitutive relation on stress-charge and strain-charge forms, respectively

$$D = e\varepsilon + \varepsilon_0 \varepsilon_{rS} E, \tag{5.30}$$

$$\varepsilon_{rS} = \begin{bmatrix} 762.5 & 0 & 0 \\ 0 & 762.5 & 0 \\ 0 & 0 & 663.2 \end{bmatrix}.$$

Piezoelectric FEM equations can be written in terms of nodal displacement $\{U\}$ and nodal electrical potential $\{\phi\}$ for each node. Mechanical efforts are equal 0, and nodal electric loads are expressed as $\{Q\}$,

$$\begin{bmatrix} [M_{uu}] & 0 \\ 0 & 0 \end{bmatrix} \begin{Bmatrix} \{\ddot{U}\} \\ \{\ddot{\phi}\} \end{Bmatrix} + \begin{bmatrix} [C_{uu}] & 0 \\ 0 & 0 \end{bmatrix} \begin{Bmatrix} \{\dot{U}\} \\ \{\dot{\phi}\} \end{Bmatrix} + \begin{bmatrix} [K_{uu}] & [K_{u\phi}] \\ [K_{u\phi}]^T & [K_{\phi\phi}] \end{bmatrix} \begin{Bmatrix} \{U\} \\ \{\phi\} \end{Bmatrix}$$
$$= \begin{Bmatrix} \{0\} \\ \{Q\} \end{Bmatrix},$$

$$\tag{5.31}$$

$$[K_{uu}] = \iiint_{\Omega e} [B_u]^T [c][B_u]dV; \qquad\qquad (5.32)$$

$$[K_{u\phi}] = \iiint_{\Omega e} [B_u]^T [e][B_u]dV; \qquad\qquad (5.33)$$

$$[K_{\phi\phi}] = \iiint_{\Omega e} [B_\phi]^T [\varepsilon][B_\phi]dV; \qquad\qquad (5.34)$$

$$[M_{uu}] = \rho \iiint_{\Omega e} [N_u]^T [N_u]dV; \qquad\qquad (5.35)$$

$$[C_{uu}] = \beta[K_{uu}], \qquad\qquad (5.36)$$

where $[K_{uu}]$—mechanical stiffness matrix, $[K_{u\phi}]$—piezoelectric coupling matrix, $[K_{\phi\phi}]$—dielectric stiffness matrix, $[M_{uu}]$—mass matrix, $\rho$—piezoelectric density, $[N_u]$—matrix of elemental shape functions, $[C_{uu}]$—mechanical damping matrix, $[B_u]$, $[B_\phi]$—derivatives of FEM shape functions, $[c]$—elastic coefficients, $[e]$—piezoelectric coefficients, $[\varepsilon]$—dielectric coefficients, $\beta$—damping coefficient, which was taken from experimental date (Table 5.13).

In the simulation piezoceramic was excited with sinusoidal (harmonic) electrical signal. The harmonic analysis allows observe the piezoelectric structure under the influence of harmonic forces, displacements, electrical charges or electrical potentials. The electrical potential was expressed by the following formula:

$$Q = A \sin(\omega), \qquad\qquad (5.37)$$

where $\omega$ is the exciting frequency.

The understanding of the resonance characteristics of vibroplatform can be investigated by visualizing the displacement, generated in the structure at resonant frequencies, which provide the resulting modal shapes. It is need to solve eigenvalue problem:

$$\begin{bmatrix} K_{UU} - \varpi C_{UU} - \varpi^2 M_{UU} & K_{U\phi} \\ K_{\phi U} & K_{\phi\phi} \end{bmatrix} \left\{ \begin{matrix} \{U\} \\ \{\phi\} \end{matrix} \right\} = \left\{ \begin{matrix} \{0\} \\ \{Q\} \end{matrix} \right\}. \qquad (5.38)$$

In order to determine the frequency of vibrations when the pressure is applied on the vibroplatform during the hot imprint process, the model was modified (Fig. 5.62).

A 20 mm × 20 mm sample of polycarbonate and the pressure tool are projected in the middle of the vibroplatform (Fig. 5.63). The pressure device is modeled as non-deformable very stiff isotropic body from structural steel. Material properties of

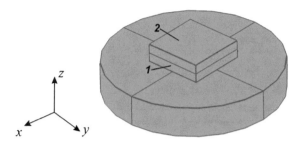

**Fig. 5.62** Modified vibroplatform with sample (*1*) and pressure tool (*2*)

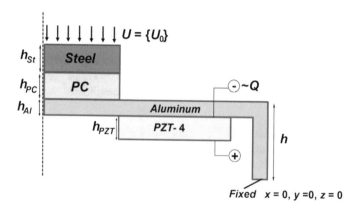

**Fig. 5.63** Computational scheme of vibroplatform with load

**Table 5.14** Materials properties

| Material properties | Polycarbonate | Structural steel |
|---|---|---|
| Young modulus (Pa) | $2 \times 10^9$ | $200 \times 10^9$ |
| Poisson's ratio | 0.37 | 0.33 |
| Density (kg/m$^3$) | 1200 | 7850 |

sample of polycarbonate and pressure tool were taken from Comsol Multiphysics library (Table 5.14). The thicknesses of steel ($h_{st}$) and polycarbonate ($h_{PC}$) are 3 mm. The pressure is defined as displacement (U) of upper steel's plane by z axis.

The full analysis consists from two steps: static analysis and eigenvalue analysis. The first step was static analysis, whose task is to find a displacement $U = \{U_0\}$ for different pressure $\{F\} = \{L, L = \overline{1,5}\, Atm\}$ as fixed load in the eigenvalue analysis.

$$[K_{UU}]\{U\} = \{F\}. \tag{5.39}$$

The second step was eigenvalue analysis:

$$\begin{bmatrix} K_{UU} - \varpi C_{UU} - \varpi^2 M_{UU} & K_{U\phi} \\ K_{\phi U} & K_{\phi\phi} \end{bmatrix} \begin{Bmatrix} \{U\} \\ \{\phi\} \end{Bmatrix} = \begin{Bmatrix} \{F(U_0)\} \\ \{Q\} \end{Bmatrix}. \tag{5.40}$$

**Experimental Research of Vibroplatform**

In the experimental research vibroplatforms, consisting from vibroplatform and piezoceramic (Fig. 5.64), of 2 mm thickness were used.

In most cases exciting frequencies of vibroplatforms are quite high, and amplitudes corresponding to them are measured in micrometers. Therefore the digital holographic interferometry system PRISM was effectively applied for the visual representation of dynamical processes, taking place in the waveguide of the vibroplatform.

PRISM system combines all the necessary equipment for deformation and vibration measurement of most materials in a small lightweight system. This system is described in Sect. 5.3.2. The results of PRISM system are images with vibration shapes at different frequencies. The vibration amplitudes were calculated analytically.

**Comparison of Simulated and Experimental Results**

From the comparison of the results between the experimental research and numerical simulation of three different vibroplatforms working in different regimes. It is clear, that simulation results correspond to experimental results of the vibroplatform of 1 mm thickness. The biggest difference was observed for the third mode of vibroplatform of 2 mm thickness. The difference of frequency is about 28% (Fig. 5.65).

Experimental research and simulation of the vibroplatform of 3 mm thickness was failed. It was impossible to identify more than the second mode. This can be explained by the lack of technical parameters of the experiment: a higher voltage was needed.

**Fig. 5.64** Vibroplatforms

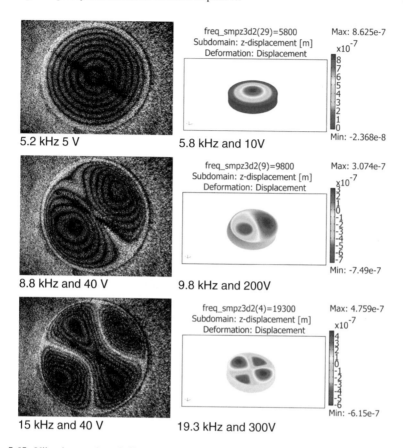

5.2 kHz 5 V                    5.8 kHz and 10V

8.8 kHz and 40 V               9.8 kHz and 200V

15 kHz and 40 V                19.3 kHz and 300V

**Fig. 5.65** Vibration modes of vibroplatform (2 mm thickness)

Summarizing, the average difference of excitation of frequencies of vibroplatform is about 10% (Table 5.15).

Another criterion of comparison is the amplitude of vibrations of the vibroplatform. On the basis of Eqs. (2.4)–(2.5) vibration amplitudes of the experimental data (Table 5.16) were calculated.

Amplitudes of simulated and analytically calculated results of vibroplatform were compared. The results are shown in Table 5.17. The average difference between vibroplatforms vibration amplitudes is about 10%.

Due to the numerical-experimental results of the vibroplatform it could be stated, that mathematical model of vibroplatform could be used for further analysis.

The next step of the investigation is to find operating frequencies of the vibroplatform, when it is under load to 5 Atm. For the development of vibroplatform, which could be applied in the mechanical hot imprint process, it is useful to

**Table 5.15** Difference of vibration frequency between experimental and simulated results

|             | Thickness of vibroplatform | | | Average (%) |
|             | 1 mm (%) | 2 mm (%) | 3 mm (%) | |
|-------------|----------|----------|----------|-------------|
| First form  | 0        | 11.5     | 5.2      | 10          |
| Second form | 8.2      | 11.4     | 1.9      |             |
| Third form  | 13.7     | 28.7     |          |             |

**Table 5.16** Vibration amplitude (nm) of experimental data

|             | Thickness of vibroplatform | | |
|             | 1 mm | 2 mm | 3 mm |
|-------------|------|------|------|
| First form  | 632  | 234  | 499  |
| Second form | 898  | 765  | 765  |
| Third form  | 499  | 233  |      |

**Table 5.17** Difference of vibration amplitudes of vibroplatform between experimental and simulated date

|             | Thickness of vibroplatform | | | Average (%) |
|             | 1 mm (%) | 2 mm (%) | 3 mm (%) | |
|-------------|----------|----------|----------|-------------|
| First form  | 19.5     | 0        | 7.0      | 10          |
| Second form | 3.9      | 2.1      | 19.6     |             |
| Third form  | 6.4      | 27.5     |          |             |

know vibration frequencies for corresponding modes under pressure of vibroplatform's surface. All these modes allow to achieve various optical properties of microstructure. The initial mathematical model of vibroplatform was modified. The pressure as constrain displacement was added to the center of the platform. Pressure parameters are applied to area of the sample (20 mm × 20 mm).

The vibration modes of vibroplatform of 2 mm thickness under the pressure of 5 Atm are presented in Fig. 5.66.

Frequencies of vibroplatforms under the pressure of 5 Atm (Table 5.18).

Table 5.18 shows that in order to get the same vibration mode, as it would be without pressure, the frequency should be increased about 20%.

### 5.3.3.2   Quality Investigation of Replicas

The mechanical hot imprint experiments where performed with periodical nickel microstructure, whose period is 4 μm. 3D view, obtained with atomic force microscope (AFM) of the original periodical microstructure is presented in Fig. 5.68a. Experiments were done at two different conditions with and without

**Fig. 5.66** Modes of 2 mm vibroplatform at 5 Atm pressure

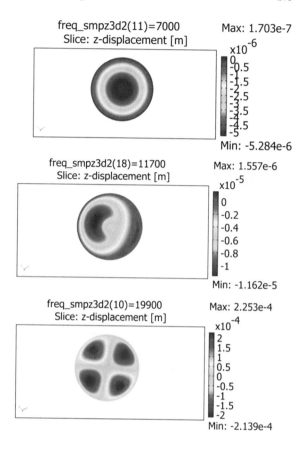

Table 5.18 Simulated frequency of vibroplatforms with pressure 5 Atm

|  | Thickness of the vibroplatform | | |
|---|---|---|---|
|  | 1 mm (kHz) | 2 mm (kHz) | 3 mm (kHz) |
| First form | 4.5 | 7 | 9 |
| Second form | 8.9 | 11.7 | 20.4 |
| Third form | 12.7 | 19.9 | |

vibration excitation (frequency 8.5 kHz, amplitude 145.6 V) at temperature of 160 °C, and pressure of 5 bar. After many experiments carried outby changing the temperature, time and frequency of vibroplatform, it was found that these process parameters have the most significant influence.

Photos, made by using optical microscope, with different scales are presented in Fig. 5.67. It is obvious that the quality of replica is better when the vibration excitation is turned on.

AFM measurements (Fig. 5.68) of the original periodical microstructure and replicas confirm the results, which were obtained by using optical microscope.

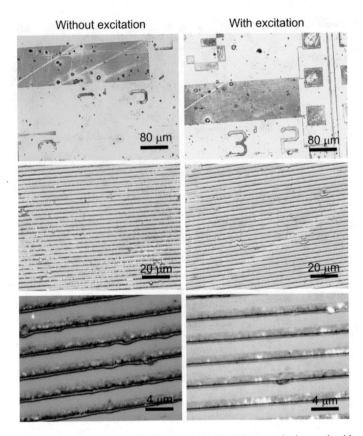

**Fig. 5.67** Photos of replicas (T = 160 °C, P = 5 Atm) without excitation and with vibration excitation (frequency 8.5 kHz, amplitude 145.6 V)

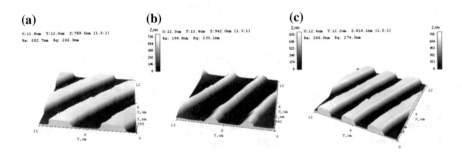

**Fig. 5.68** 3D view of the original periodical microstructure (**a**) and its replicas done without excitation (**b**) and with vibration excitation (**c**)

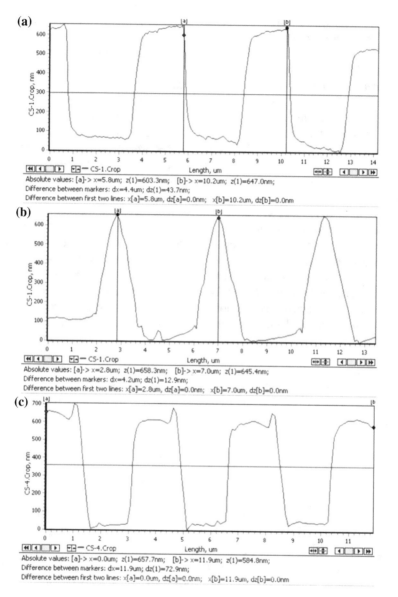

**Fig. 5.69** The profile view of the original periodical microstructure (**a**) and its replicas done without excitation (**b**) and with vibration excitation (**c**)

Replicas done with vibration excitation look like the original structure (Fig. 5.68). The average depth of the original stamp is 567.65 nm, the average depth of the replica, obtained without excitation is 545.36 nm, and the average depth of the replica produced with vibration excitation is 556.69 nm. High frequency vibrations help to fill gaps of the original structure with polymer. The filling of gaps increases from 62.5 to 75% (Fig. 5.69). The surface roughness is another important parameter, which strongly influences optical, electrical and mechanical properties of the replicated structure. Vibration excitation definitely helps to improve the quality of the surface of replicas. Roughness measurements of the surface of the original stamp and replicas are shown in Fig. 5.70. The surface roughness of nickel stamp is approximately 8 μm, whereas replicas' surface roughness is much higher: 55 μm is the roughness of the replica, which was produced without vibration excitation and 23 μm is the roughness of the replica produced with vibration excitation (Fig. 5.71). The roughness of replicas is still much larger compared to the original stamp, but the vibration excitation helps to decrease it about 50%.

The quality of replicas was tested using an indirect optical method—this method is based on the measurement of the diffraction efficiency. Replicated periodical structures are manufactured from optical materials, so optical methods can be used for the evaluation of the quality of replicas. Diffraction efficiency of i-th maximum was measured using a photodiode and calculated as the ratio of the energy of its maximum and total energy reflected from the analyzed microstructure. This method of the diffraction efficiency calculation allows eliminate properties of the material of the microstructure and helps analyze and compare its geometrical properties.

Measurements show, that high frequency vibration excitation increases the diffraction efficiency of the first order maximum 1.5 times (from 11.25% measured for the microstructure replicated without excitation up to 17.89% for the microstructure replicated with vibration excitation) (Fig. 5.72). It can be noted that this is still by 40% worse than the theoretical result (32%), but it shows a promising direction for the future research.

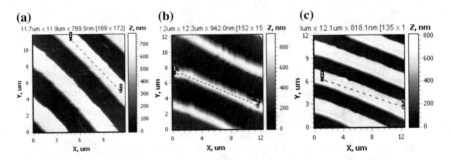

**Fig. 5.70** 2D view and roughness measurement of the original periodical microstructure (**a**) and its replicas produced without excitation (**b**) and with vibration excitation (**c**)

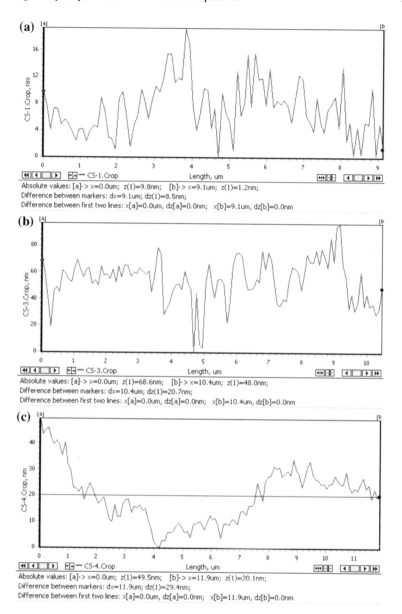

**Fig. 5.71** Ridge roughness of the original periodical microstructure (**a**) and its replicas done without excitation (**b**) and with vibration excitation (**c**)

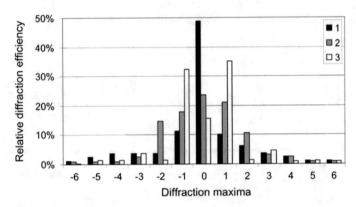

**Fig. 5.72** Relative diffraction efficiencies of replicas produced without excitation (*1*) and with vibration excitation (*2*) and relative diffraction efficiencies of the ideal replica (*3*) (profile— lamellar, period—4 μm, land—2 μm, ridge—2 μm, depth—560 μm)

# References

1. Bruck HA, Yang M, Kostov Y, Rasooly A (2013) Electrical percolation based biosensors. Methods 63:282–289
2. Ziegler C (2004) Cantilever-based biosensors. Anal Bioanal Chem 379:946–959
3. Tomkins MR, Chow J, Lai Y, Docoslis AA (2013) Coupled cantilever–microelectrode biosensor for enhanced pathogen detection. Sens Actuators B Chem 176:248–252
4. Bausells J (2015) Piezoresistive cantilevers for nanomechanical sensing. Microelectron Eng 145:9–20
5. Tran AT, Wunnicke O, Pandraud G, Nguyen MD, Schellevis H, Sarro PM (2013) Slender piezoelectric cantilevers of high quality AlN layers sputtered on Ti thin film for MEMS actuators. Sens. Actuators A Phys. 202:118–123
6. Wang J, Chen D, Xu Y, Liu W (2014) Label-free immunosensor based on micromachined bulk acoustic resonator for the detection of trace pesticide residues. Sens Actuators B Chem 190:378–383
7. Medjahdi N, Benmoussa N, Benyoucef B (2014) Modeling, simulation and optimization of the mechanical response of micromechanical silicon cantilever: application to piezoresistive force sensor. Phys Procedia 55:348–355
8. Stolyarova S, Cherian S, Raiteri R, Zeravik J, Skladal P, Nemirovsky Y (2008) Composite porous silicon-crystalline silicon cantilevers for enhanced biosensing. Sens Actuators B Chem 131:509–515
9. Sappia LD, Trujillo MR, Lorite I, Madrid RE, Tirado M, Comedi D, Esquinazi P (2015) Nanostructured ZnO films: a study of molecular influence on transport properties by impedance spectroscopy. Mater Sci Eng B 200:124–131
10. Khun K, Ibupoto ZH, Chey CO, Lu J, Nur O, Willander M (2013) Comparative study of ZnO nanorods and thin films for chemical and bio-sensing applications and the development of ZnO nanorods based potentiometric strontium ion sensor. Appl Surf Sci 268:37–43
11. Kakimoto K, Fukata K, Ogawa H (2013) Fabrication of fibrous BaTiO$_3$-reinforced PVDF composite sheet for transducer application. Sens Actuators A Phys 200:21–25
12. Pérez R, Král M, Bleuler H (2012) Study of polyvinylidene fluoride (PVDF) based bimorph actuators for laser scanning actuation at kHz frequency range. Sens Actuators A Phys 183:84–94

13. Xu L, Ling SF, Lu B, Li H, Hu H (2006) Sensing capability of a PZT-driven cantilever actuator. Sens Actuators A Phys. 127:1–8
14. Liu H, Quan C, Tay CJ, Kobayashi T, Lee C (2011) A MEMS-based piezoelectric cantilever patterned with PZT thin film array for harvesting energy from low frequency vibrations. Phys Procedia 19:129–133
15. Albert J, Lepinay S, Caucheteur C, DeRosa MC (2013) High resolution grating-assisted surface plasmon resonance fiber optic aptasensor. Methods **63**:239–254
16. Tripathi SM, Bock WJ, Mikulic P, Chinnappan R, Ng A, Tolba M, Zourob M (2012) Long period grating based biosensor for the detection of *Escherichia coli* bacteria. Biosens Bioelectron 35:308–312
17. Fan X, White IM, Shopova SI, Zhu H, Suter JD, Sun Y (2008) Sensitive optical biosensors for unlabeled targets. Anal Chim Acta 620:8–26
18. Piliarik M, Homola J (2009) Surface plasmon resonance (SPR) sensors: approaching their limits? Opt Express 17:16505–16517
19. Bhatta D, Stadden E, Hashem E, Emmerson G (2010) Multi-purpose optical biosensors for real-time detection of bacteria, viruses and toxins. Sens Actuators B 149:233–238
20. Miroslav P, Petr S, Michal K (2007) Biosensors for biological warfare agent detection. Defence Sci J 57:185–193
21. Gerchberg RW, Saxton WO (1972) A practical algorithm for the determination of the phase from image and diffraction plane pictures. Optik 35:237–246
22. Dufresne ER, Spalding GC, Dearing MT, Sheets SA, Grier DG (2001) Computer-generated holographic optical tweezer arrays. Rev Sci Instrum 72(3):1810–1816
23. Deng S, Liu L, Lang H, Zhao D (2006) Hiding an image in cascaded Fresnel digital holograms. Chin Optics Lett 4(5):268–271
24. Yujun F, Ding J, Jin Z, Xinfan H, Wenqi G (2000) Self-focusing matched filter produced by computer-generated hologram. Optics Commun 184(1–4):89–93
25. Ferchera AF (2010) Computer-generated holograms for testing optical elements: error analysis and error compensation. Optica Acta Int J Optics 23(5):347–365
26. Liu J, Jin H, Ma L, Jin W (2013) Optical color image encryption based on computer generated hologram and chaotic theory. Optics Commun 307:76–79
27. Palevicius P, Ragulskis M (2014) Image communication scheme based on dynamic visual cryptography and computer generated holography. Optics Communications 335:161–167
28. Liu C, Yang Y, Guo S, Xu R, Men T, Wen C (2013) Modal wavefront sensor employing stratified computer-generated holographic elements. Opt Lasers Eng 51(11):1265–1271
29. Liu C, Men T, Xu R, Wen C (2014) Analysis and demonstration of multiplexed phase computer-generated hologram for modal wavefront sensing. Optik Int J Light Electron Optics 125(11):2602–2607
30. Ozols A, Augustovs P, Saharov D (2012) Recording of holographic gratings and their coherent self-enhancement in AN a-As2S3 film with a minimum light intensity modulation. Lith J Phys 52:10–18
31. Worgull M (2010) Hot embossing. Micromanuf Eng Technol 68–89
32. Gale MT et al (2005) Replication technology for optical Microsystems. Opt Lasers Eng 43:373–386
33. Calixto S et al (2007) Fabrication of transmissive diffractive optical elements for the mid-infrared with a laser writing instrument. J Appl Res Technol 5(2):74–88
34. Poleshchuk AG et al (2002) Methods for certification of CGH fabrication. In: Proceedings OSA/DOMO, pp 438–440
35. Freese W et al (2011) Optimized electron beam writing strategy for fabricating computer-generated holograms based on an effective medium approach. Opt Express 19 (9):8684–8692
36. Gao F et al (2002) Electron-beam lithography to improve quality of computer-generated hologram. Microelectron Eng 61–62:363–369

37. Pang YK et al (2005) Chiral microstructures (spirals) fabrication by holographic lithography. Opt Express 13(19):7615–7620
38. Barbero DR et al (2007) High resolution nanoimprinting with a robust and reusable polymer mold. Adv Funct Mater 17:2419–2425
39. Ito H et al (2009) Polymer structure and properties in micro- and nanomolding process. Curr Appl Phys 9:19–24
40. Lee KB (2011) Principles of microelectromechanical systems. Wiley, New York, p 667
41. Minguez-Vega G et al (2007) Dispersion-compensated beam-splitting of femtosecond light pulses: Wave optics analysis. Opt Express 15:278–288
42. Liu J et al (2004) Design of diffractive optical elements for high-power laser applications. Opt Eng 43:2541–2548
43. Emiliani V et al (2005) Wave front engineering for microscopy of living cells. Opt Express 13:1395–1405
44. Jureller JE, Kim HY, Scherer NF (2006) Stochastic scanning multiphoton multifocal microscopy. Opt Express 14:3406–3414
45. Yang XL et al (2003) Interference of four umbrella like beams by a diffractive beam splitter for fabrication of two-dimensional square and trigonal lattices. Opt Lett 28:453–455
46. Ferraro P et al (2005) Extended focused image in microscopy by digital holography. Opt Express 13:6738–6750
47. Mann CJ, Yu L, Kim MK (2006) Movies of cellular and sub-cellular motion by digital holographic microscopy. Biomed Eng Online 23:5–21
48. Satake S et al (2007) Parallel computing of a digital hologram and particle searching for microdigital-holographic particle-tracking velocimetr. Appl Opt 46:538–550
49. Charriere F et al (2006) Characterization of microlenses by digital holographic microscopy. Appl Opt 45:829–837
50. Horimai H, Tan X (2006) Collinear technology for a holographic versatile disk. Appl Opt 45:910–922
51. Poon TH (2006) Digital holography and three-dimensional display: principles and applications. Springer, Berlin, 430 p
52. Yasuda T et al (2009) Computer simulation of reconstructed image for computer-generated holograms. In: Proceedings of the SPIE, vol 7233, p 72330H-11
53. Gerchberg RW, Saxton WO (1972) A practical algorithm for the determination of phase from image and diffraction plane pictures. Optik 35:227–246
54. Sakamoto Y, Nagao TA (2002) Fast computational method for computer-generated Fourier hologram using patch model. Electron Commun Japan 85(11):16–24
55. Tamura H, Ishii Y (2012) Computer-generated hologram fabricated by electron-beam lithography for noise reduction. Opt Rev 19(2):50–57
56. Tamura H, Torii Y (2012) Enhancement of the Lohmann-type computer-generated hologram encoded by direct multilevel search algorithm. Opt Rev 19(3):131–141
57. Oikawa M et al (2011) Computer-generated hologram using an approximate Fresnel integral. J Opt 13(7):075405
58. Petruškevičius R et al (2010) E-beam lithography of computer generated holograms using a fully vectorial 3D beam propagation method. Microelectron Eng 87:2332–2337
59. Brigham EO (1988) Fast Fourier transform and its applications. Prentice-Hall, Prentice
60. Thomson MJ, Taghizadeh MR (2005) Design and fabrication of Fourier plane diffractive optical elements for high-power fibre-coupling applications. Opt Lasers Eng 43:671–681
61. Mihailescu M et al (2007) Optimization of the reconstruction parameters in computer generated holograms and digital holography. U.P.B. Sci Bull Series A 69(3):25–36
62. O'Shea DC et al (2004) Diffractive optics—design, fabrication, and test. Spie Press, Bellingham, p 15
63. Caballero JAD et al (2009) Design and fabrication of computer generated holograms for fresnel domain lithography. In: OSA DH and 3D Imaging, 2009, p DWB3

64. Andrulevičius M, Tamulevičius T, Tamulevičius S (2007) Formation and analysis of dot-matrix holograms. Mater Sci (Medžiagotyra) 13(4):278–281
65. Andrulevičius M et al (2010) Hologram origination combining rainbow and dot-matrix holograms. Mater Sci (Medžiagotyra) 16(4):298–301
66. Abd Rahman SF et al (2011) Fabrication of nano and micrometer structures using electron beam and optical mixed lithography process. Int J Nanoelectron Mater 4:49–58
67. Wilson WL et al (1998) High density, high performance optical data storage via volume holography: viability at last? Opt Quant Electron 32:393–404
68. Gren P (2003) Four-pulse interferometric recordings of transient events by pulsed TV holography. Opt Lasers Eng 40:517–528
69. Saari P, Kaarli R, Ratsep M (1992) Temporally multiplexed fourier holography and pattern recognition of femtosecond-duration images. J Lumin 56:175–180
70. Nishchal NK, Joseph J, Singh K (2004) Fully phase encryption using digital holography. Opt Eng 43:2959–2966
71. Son JY, Javidi B, Kwack KD (2006) Methods for displaying three-dimensional images. Proc IEEE 94:502–523
72. Короленко ПВ (1997) Оптика когерентного излучения, 222 p
73. Короленко ПВ (1998) Оптика коге`рентного излучения. Московский Университет, 155 p
74. Whyte G, Courtial J (2005) Experimental demonstration of holographic three-dimensional light shaping using a Gerchberg-Saxton algorithm. New J Phys 7:117–128
75. Heckele M, Schomburg WK (2004) Review on micro molding of thermoplastic polymers. J Micromech Microeng 14:1–14
76. Worgull M et al (2006) Modeling and optimization of the hot embossing process for micro- and nanocomponent fabrication. Microsyst Technol 12:947–952
77. Wu CH, Kuo HC (2007) Parametric study of injection molding and hot embossing polymer microfabrication. J Mech Sci Technol 21:1477–1482
78. Becker H, Gartner C (2000) Polymer microfabrication methods for microfluidic analytical applications. Electrophoresis 21:12–26
79. Lee LJ et al (2001) Design and fabrication of CD-like microfluidic platforms for diagnostics: Polymer based microfabrication. Biomed Microdevice 3(4):339–351
80. Krauss PR, Chou SY (1997) Nano-compact disks with 400 Gbit/in 2 storage density fabricated using nanoimprint lithography and read with proximal probe. Appl Phys Lett 71:3174–3176
81. Jaszewski RW et al (1998) Hot embossing in polymers as a direct way to pattern resist. Microelectron Eng 41–42:575–578
82. Juang YJ (2001) Polymer processing and rheological analysis near the glass transition temperature. Dissertation, 230 p
83. Li JM, Liu C, Peng J (2008) Effect of hot embossing process parameters on polymer flow and microchannel accuracy produced without vacuum. J Mater Process Technol 207: 163–171
84. Giboz J, Copponex T, Mele P (2007) Microinjection moulding of thermoplastic polymers: a review. J Micromech Microeng 17:96–109
85. Sahli M et al (2009) Quality assessment of polymer replication by hot embossing and micro-injection moulding processes using scanning mechanical microscopy. J Mater Process Technol 209:5851–5861
86. Lin CR, Chen RH, Chen CH (2003) Preventing non-uniform shrinkage in open-die hot embossing of PMMA microstructures. J Mater Process Technol 140:173–178
87. Yoshihiko H, Yoshida S, Nobuyuki T (2003) Defect analysis in thermal nanoimprint lithography. J Vacuum Sci Technol B 21:2765–2770
88. Hu XF (2005) Research on the equipment for hot embossing and related experiments. Master thesis

89. Yao DG, Vinayshankar LV, Byung K (2005) Study on sequeezing flow during nonisothermal embossing of polymer microstructure. J Polym Eng Sci 45:652–660
90. Liu C et al (2010) Deformation behavior of solid polymer during hot embossing process. Microelectron Eng 87:200–207
91. Mekaru H, Goto H, Takahashi M (2007) Development of ultrasonic micro hot embossing technology. Microelectron Eng 84:1282–1287
92. Hirai Y, Yoshida S, Takagi N (2003) Defect analysis in thermal nanoimprint lithography. J Vac Sci Technol, B 21(6):2765–2770
93. Narijauskaitė B et al (2011) High-frequency excitation for thermal imprint of microstructures into a polymer. Experimental Techniques. (published online: 11 APR 2011)
94. Seunarine K et al (2006) Optical heating for short hot embossing cycle times. Microelectron Eng 83:859–863
95. Liu JS, Dung YT (2005) Hot Embossing Precise Structure Onto Plastic Plates by Ultrasonic Vibration. Polym Eng Sci 45:915–925
96. Yule AJ, Al-Suleimani Y (2000) On droplet formation from capillary waves on a vibrating surface. Math Phys Eng Sci 456:1069–1085
97. Zhang A et al (2009) Rapid concentration of particle and bioparticle suspension based on surface acoustic wave. Appl Acoust 70(8):1137–1142
98. Kondoh J et al (2009) Development of temperature-control system for liquid droplet using surface Acoustic wave devices. Sens Actuators, A 149(2):292–297
99. Lomonosov AM, Hess P (2008) Nonlinear surface acoustic waves: realization of solitary pulses and fracture. Ultrasonics 48(6–7):482–487
100. Alvarez M, Friend JR, Yeo LY (2008) Surface Vibration induced spatial ordering of periodic polymer patterns on a substrate. Langmuir 24:10629–10632
101. Changliang X, Mengli W (2005) Stability analysis of the rotor of ultrasonic motor driving fluid directly. Ultrasonics 43(7):596–601
102. Ming Y et al (2005) Design and evaluation of linear ultrasonic motors for a cardiac compression assist device. Sens Actuators, A 119(1):214–220
103. Nakamura K, Maruyama M, Ueha S (1996) A new ultrasonic motor using electro-rheological fluid and torsional vibration. Ultrasonics 34(2–5):261–264
104. Fairbanks HV (1974) Applying ultrasonics to the moulding of plastic powders. Ultrasonics 12:22–24
105. Paul DW, Crawford RJ (1981) Ultrasonic moulding of plastic powders. Ultrasonics 19: 23–27
106. Khuntontong P, Blaser, T, Schomburg WK (2008) Ultrasonic micro hot embossing of polymer exemplified by a micro thermal flow sensor. In: Proceedings of smart system integration, Barcelona, Spain, 9–10 April 2008, pp 327–334
107. Benatar A, Gutowski TG (1989) Ultrasonic welding of PEEK graphite APC-2 composites. Polym Eng Sci 29(23):1705–1721
108. Nonhof CJ, Luiten GA (1996) Estimates for process conditions during the ultrasonic welding of thermoplastics. Polym Eng Sci 36:1177–1183
109. Jaffe B, Cook WR, Jaffe H (1971) Piezoelectric Ceramics. Academic, New York
110. Вест Ч (1982) Голографическая интерферометрия. Москва: Мир
111. McCaslin SE et al (2012) Closed-form stiffness matrices for higher order tetrahedral finite elements. Adv Eng Softw 44(1):75–79

Printed in the United States
By Bookmasters